EDUCATED GUESSING

POPULAR STATISTICS

a series edited by

D. B. Owen
Department of Statistics
Southern Methodist University
Dallas, Texas

Nancy R. Mann
Biomathematics Department
UCLA
Los Angeles, California

Other Volumes in Preparation

EDUCATED GUESSING

How to Cope in an Uncertain World

SAMUEL KOTZ
University of Maryland
College Park, Maryland

DONNA F. STROUP
University of Texas
Austin, Texas

MARCEL DEKKER, INC.   New York and Basel

Library of Congress Cataloging in Publication Data
Kotz, Samuel.
   Educated guessing.

   (Popular statistics ; 2)
   Bibliography: p.
   Includes index.
   1. Probabilities. 2. Statistics. I. Stroup,
Donna F., [date]. II. Title.
QA273.K664   1983   519.2   83-7222
ISBN 0-8247-7000-5

MARCEL DEKKER, INC.
270 Madison Avenue, New York, New York 10016

Current printing (last digit):
10 9 8 7 6 5 4 3 2 1

PRINTED IN THE UNITED STATES OF AMERICA

PREFACE

Probably you are reading this book because you expect to be informed about the notions of *probability, expectation,* and *information*. At least, that is our guess. Now look at that first sentence again. The first uses of the words *probably, expect*, and *informed* come from colloquial language. A native English speaker would have learned the meaning of these terms sometime during childhood, while those for whom English is a second language understand the words as a variety of terms. These words have interpretations varying from person to person and from region to region, and yet they have enough common meaning that we can write them down and expect you to understand the sentence. The words *probability, expectation,* and *information* have precise interpretations in the context of our study. The purposes of this volume are to bridge the gap between vague intuitive meanings and precise mathematical formulations and to show how a clear understanding of the concepts can be utilized in various aspects of the real world: scientific research, management decisions, recreational sports, (educated) gambling, political involvement, or ordinary interactions with people. One of our aims is to combat mathematical anxiety, which is magnified in dealing with chance, and to emphasize the relevance of probability theory in communication and decision making.

Probability theory originated with games of chance, but in recent years this connection has been avoided in probability texts. However, gambling persists in becoming more widespread than ever. Uneducated gambling can result in devastating consequences; educated gambling is the best prevention. Before visiting a casino—if you must—we suggest a careful reading of this book. Our work is an outgrowth of the popular course The Art of Guessing taught in the Department of Mathematics of Temple University continuously since 1975. The central chapter, Chapter 3, "Controlling Uncertainty," no doubt was influenced by events leading to the legalization of casino gambling in Atlantic

City, New Jersey, in May 1978. Chapter 4, "Information Theory," is useful for specialists in communications, computer science, and psychology. Chapter 5, "Decision Making in Uncertainty," is a response to widespread introduction of quantitative methods in business and management. Appendix A, "Paradoxes of Probability," is addressed to those interested in logic, philosophy, and fun with problem solving in general.

We are deeply indebted to many people who have contributed to the production of this text: Professor Roy Kuebler, Jr., whose experience and mastery substantially formed the first two chapters; Professor G. P. Patil, whose encouragement and assistance provided many wise and helpful suggestions; Professor Bonnie Auerbach, whose comments and suggestions for revisions were invaluable; Professor W. L. Harkness, for support, encouragement, and advice; the late Professor J. Wolfowitz, Professor R. J. Larsen, and Dr. Henry I. Braun, our instructors of probability, who taught us with their lucid and informative lectures; our students, who taught us to understand the importance of probability theory in human activity; the helpful staff of the libraries of Temple University, the University of North Carolina at Chapel Hill, Pennsylvania State University, and the University of Texas at Austin; the editors, staff, and reviewers of Marcel Dekker, Inc., for the endless transformation of a manuscript into a published book, and especially Dr. Maurits Dekker, Chairman of the Board, for his enthusiastic support and invaluable guidance in this work; our families, for their patience, understanding, and appreciation of our endeavor.

Samuel Kotz
Donna F. Stroup

CONTENTS

EDUCATED GUESSING

1

ASSESSING PROBABILITIES

1.1 INTRODUCTION

A pinch of probably is worth a pound of perhaps.
James Thurber

We as human beings strive to exert an increasing amount of influence over our own world: we build a nuclear reactor, we conserve natural gas, we dump wastes, we invent cures, we endanger wildlife. Our tendency is to deny the presence of uncertainty, but we are often unsuccessful. An investor must choose between buying stocks and tying up assets in real estate in an uncertain economic period. A juror must cast a vote, guilty or innocent, in the trial of a defendant. A college student must decide how much time should be spent studying for that midterm exam. A gambler must decide when to pack up the winnings and head home.

At the center of all these problems is the concept of probability. We use this idea commonly in our colloquial language. "It will *probably* rain next Monday, since I am planning a fishing trip." "The Yankees will *probably* not make it to the World Series this year." We may also use more quantitative expressions. "The *probability* a coin lands heads up is 1/2." "The *probability* of rain tomorrow is 40 percent."

In this discussion, we must distinguish between the adverb *probably* in the first two sentences and the noun *probability* in the last two. The difference lies in the kind of situation involved and in our assessment of it. The tossing of a coin or the outcome of tomorrow's weather is an uncertain operation in which our stated outcome (head, rain) is one of a number of possible outcomes, none of which is certain to occur. We assess the likelihood of the stated outcome by comparing it with all of its competitors. If we think that the coin and its toss are both "fair," then we argue that the two outcomes "head" and "tail" are each equally likely to occur. We then express the probability of "head" as 1/2 since "head" is one of two equally likely outcomes. In the case of the weather, meteorological records have indicated that in 40 percent of the cases in which weather conditions were similar to today's, rain followed.

Any such uncertain operation can, in principle, be repeated indefinitely often under identical circumstances, and so our assessment of probability carries with it the notion of relative frequency in the long run. In the long run of coin tosses, we anticipate that 1/2 of the total number of tosses will show "head." In a long run of weather records, 40 percent of all days like today will be followed by rain.

In the adverbial cases, when we say "It will probably rain next Monday" or "The Yankees will probably not make it to the World Series this year," we are referring to a single situation which will not be repeated again. Our use of *probably* expressed only some vague degree of belief that the stated event will occur. There is, of course, no reason why a person should not make a degree of belief numerically precise if that suits a purpose. After all, Lloyd's of London has the reputation of being willing to insure against any kind of single-shot hazard for a specific insurance premium which they will quote. Forced to bet on the Yankees' participation in the World Series, we shall have to qualify our degree of belief at least to the extent of specifying how much we should win if they do play and how much we should lose if they do not. Any such qualification leads to a number that acts like a probability.

This duality in the concept of probability was recognized as early as the seventeenth century, when the popularity of games of chance spawned interest in probability, at least among the upper classes in France and England. However, in the realm of gambling the two meanings of probably seem to coincide. We say that the chances of getting two aces (one dot on a face) in a row when casting a die are slim (or the probability is low). We know this because we can calculate numerically this probability to be 1/36, which is felt to be a small number close to zero. On the other hand, this event is also improbable in the subjective sense because we really do not believe very strongly that the next two tosses of the die will result in aces.

Historically, many mathematicians have emphasized the objective view of probability, organizing a general structure to explain uncertain situations and give rules of probability as guides for action. However, in recent years debate about the foundations of probability has become lively; a strong body of research holds that the notion of objective probabilities is unnecessary, even improper, and that the probability of any proposition must be evaluated subjectively (for example, Bruno de Finetti, 1974). If experiments can be performed to narrow the range of uncertainty, then they should be. But when experimental evidence is not available, decisions must be made on the basis of the opinions of uncertain experts. Dickey (1979) proposes that we view probability as a language, "a tool for expression, for articulation of thought. Like any tool, it has an effect on its user, helps mold thought as well as communicate it." In this chapter we invite you to resist the human tendency to avoid uncertainty by investigating this language of probability.

1.2 PROBABILITY THEN AND NOW

How dare one speak of the laws of chance? Is not chance the very antithesis of all law?

Joseph Bertrand

Although we use the words *subjective* or *objective* to describe probabilities, we are careful not to use the word *arbitrary*. Bertrand answered his own questions by extended exposition of the laws which may be said to describe the operation of uncertainty. It is not that the outcome of an uncertain experiment is certain or deterministic, but rather that we can uncover a stable pattern to the various possible uncertain outcomes. As he wrote subsequently, "Chance at each trial counteracts its caprices. Even irregularities have their law."

To investigate these "laws of irregularities," let us consider the toss of a balanced (fair) coin. The outcome of such a trial is uncertain, but we know it will be either "head" or "tail." We say that these two outcomes *exhaust* the possibilities. Furthermore, these two outcomes are *mutually exclusive*, meaning that if one occurs, the other cannot. And if the coin is balanced and the toss is fair, the two outcomes will be *equally likely*. To express the probability of getting "head" in the toss of a coin, we appeal to the following classical definition of probability.

> **Probability: Classical Definition** If a single trial of a chance situation can have one of N different outcomes which are exhaustive, mutually exclusive, and equally likely, and if f of these N possibilities are favorable to a specified event A, then the probability of the event A, written P(A), is defined as f/N.

Let us return to our example of the coin toss; N = 2, and of these two outcomes exactly one is favorable to getting "head." Consequently, f = 1 and P(head) = 1/2. If we roll a six-sided die, the face on top when the die comes to rest will show 1, 2, 3, 4, 5, or 6. These six outcomes are exhaustive and mutually exclusive; they are equally likely if the die is perfectly balanced. Thus, N = 6. The probability of any one of these outcomes is 1/6 since only one outcome is favorable for any specified number. If the event of interest is "a number divisible by 3," then there are two of the six possible outcomes favorable to this event, 3 and 6. Hence

$$P(\text{number divisible by 3}) = \frac{2}{6} = \frac{1}{3}$$

Let us roll two dice and add the numbers of dots shown on both resulting up-faces, calling the sum the point which has been rolled. This point will be some number between 2 and 12, eleven outcomes which are exhaustive and mutually exclusive but not equally likely. To illustrate, let us agree to indicate a possible pair of faces of the two dice by writing (x, y), where x indicates the reading on the first die and y indicates the reading on the second die. Then the 36 possible pairs are equally likely:

(1, 1)	(2, 1)	(3, 1)	(4, 1)	(5, 1)	(6, 1)
(1, 2)	(2, 2)	(3, 2)	(4, 2)	(5, 2)	(6, 2)
(1, 3)	(2, 3)	(3, 3)	(4, 3)	(5, 3)	(6, 3)
(1, 4)	(2, 4)	(3, 4)	(4, 4)	(5, 4)	(6, 4)
(1, 5)	(2, 5)	(3, 5)	(4, 5)	(5, 5)	(6, 5)
(1, 6)	(2, 6)	(3, 6)	(4, 6)	(5, 6)	(6, 6)

What is the probability of rolling the point 7? Out of these 36 exhaustive, mutually exclusive, and equally likely outcomes, the ones favorable to point 7 are:

$$(1, 6) \quad (2, 5) \quad (3, 4) \quad (4, 3) \quad (5, 2) \quad (6, 1)$$

Thus $f = 6$ and $P(\text{point } 7) = 6/36 = 1/6$.

Now let us consider an uncertain situation where the number of outcomes is infinite: we shall toss a coin until the first head appears. Using H for head and T for tail, then the possible outcomes of a trial for this uncertain experiment are:

H	(head on first toss)
TH	(head on second toss)
TTH	(head on third toss)
TTTH	(head on fourth toss)
TTTTH	(head on fifth toss)

$$\begin{matrix} . & & . \\ . & & . \\ . & & . \end{matrix}$$

and so on forever. There is no direct way to calculate f/N here for any event of interest.

We may resolve some of the logical difficulties by replacing the f/N definition by a procedure of assigning to every possible outcome a number between 0 and 1, with the restriction that the sum of these numbers for all possible outcomes must add up to 1. The number attached to an outcome is its *probability mass*, and the probability of an event is simply the sum of the probability masses of all outcomes favorable to the event. In the case of rolling a pair of dice, we assign mass 1/36 to each of the 36 possible outcomes listed above. Then

P(point 7) = P[(1, 6), (2, 5), (3, 4), (4, 3), (5, 2), (6, 1)] = 1/36 + 1/36 + 1/36 + 1/36 + 1/36 + 1/36 = 6/36 = 1/6.

To illustrate the pitfalls we encounter when assessing probabilities according to the classical definition, consider the following situation.

Example 1.2.1 Two people are to be selected as cochairpersons from a group of two women and two men. What is the probability that the two chairpersons are of the same sex?

Pitfall I. There are three cases: two female chairpersons, two male chairpersons, or chairpersons of different sexes. In two cases of the three the sexes match; therefore the matching probability is 2/3. (This pitfall could have been avoided by recognizing that the three cases are not equally likely.)

Pitfall II. There are four cases: two female chairpersons, two male chairpersons, female chairperson x and male chairperson y, and male chairperson x and female chairperson y. In two of the four cases the sexes match. Therefore, the matching probability is 1/2. (These cases are not equally likely either.)

Correct Solution. If the four people in the group are denoted F_1, F_2, M_1, and M_2, then there are six equally likely ways of choosing two people out of four: (F_1, F_2), (F_1, M_1), (F_1, M_2), (F_2, M_1), (F_2, M_2), and (M_1, M_2). Out of these six cases only two are pairs of people of the same sex. Hence the matching probability is 1/3.

Exercises

1.1   In the roll of a single die, what is the probability that the resulting number will be
       a.   A prime number (that is, a number divisible only by itself and 1)?
       b.   An even number?

1.2   In the text we discussed the fair roll of a pair of fair dice and found the probability that the point rolled would be 7. Calculate the probability of each of the other possible points and confirm the results of Table B.1 in Appendix B.

1.3   A deck of playing cards for bridge or poker contains 52 cards, composed of 13 denominations (2, 3, 4, 5, 6, 7, 8, 9, 10, jack, queen, king, ace) and 4 suits (clubs, diamonds, hearts, spades). Clubs and spades are black in color; diamonds and hearts are red. When the deck has been well shuffled and a card is drawn from it at random, the conditions for equally likely outcomes have been satisfied. What is the probability that a single card drawn will be
       a.   A spade?
       b.   A black card?

      c. A face card (jack, queen, or king)?

      d. An honor card (ace, king, queen, jack, ten)?

1.4   What is the probability of tossing

      a. Two heads in the toss of two coins?

      b. Two heads in the toss of three coins?

1.5   In an ancient matching game, two players show their right hands to each other simultaneously, with one, two, or three fingers extended. If each player is equally likely to show one, two, or three fingers, what is the probability that the total number of fingers showing is

      a. Odd?

      b. Even?

      c. Less than 4?

      d. Either the largest or the smallest value possible?

1.6   A parking lot contains only two white cars and one black car. If two people arrive at random to claim their cars, what is the probability that they will claim cars of different colors?

1.7   If an academic department is composed of two women and three men, what is the probability that two people chosen at random to serve on a curriculum committee will be of the same sex?

1.8   If three dice are rolled together, what is the probability that the point (sum) will be 5? (Hint: The total number of outcomes such as (1, 3, 5), etc., will be (6) (6) (6) = 216.)

1.9   In rolling three fair dice, show that 1/8 is the probability for a total of 10 and also for a total of 11. Show that 10 and 11 are the most likely of all outcomes.

1.10   In a knockout tournament of size 4, the players are subdivided randomly into two groups of two players each. In the first round of the tournament, the first drawn player of the first group plays the first drawn player of the second group, and the second drawn players of the two groups play each other. The winners in the first round are then paired in the second round. The winner of the second round is the winner of the tournament and his or her opponent is the first runner-up. Assume that in every contest the better player wins. What is the probability that

      a. The best player of the four will be the winner of the tournament?

      b. The second-best player will be the first runner-up?

      c. The third-best player will be the first runner-up?

      d. The weakest player will be the first runner-up?

1.3 THE LAW OF AVERAGES

Instead of thus appealing to the proportion of cases favorable to the event, it is far better . . . to appeal at once to the proportion of cases in which the event actually occurs.

                                              John Venn

Venn is credited as the first to give a precise and consistent definition of probability in terms of actual experience. He argued that the terms *probability* and *chance* presuppose a series of trials, some of which exhibit an attribute of interest. The probability of interest is then defined as the numerical fraction which represents the proportion between the two different classes in the long run. This definition of probability is of great help to statisticians who rely on our subject. Consider the cumulative experience obtained by repeating an experiment many, many times. For example, let's toss our coin and after each toss calculate what fraction of all of our tosses came up heads. Figure 1.1 is a picture of the first 100 of such an experiment; you may want to draw your own picture for a smaller number of tosses. Intuitively, it seems that if we toss the coin many thousands of times, the graph will settle down to the horizontal line, a constant; that is, the variation in the fraction of heads becomes negligible. It is this constant that we call the probability of a head:

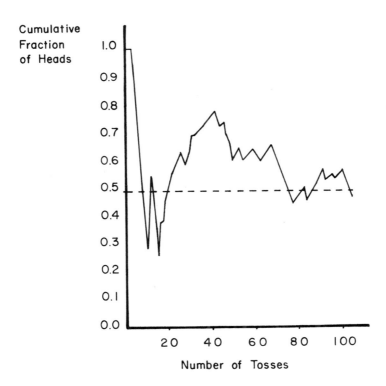

Figure 1.1   Fraction of heads in repeated tosses of a fair coin.

$$P(H) = \lim_{n \to \infty} \left(\frac{\text{number of heads in n tosses}}{n} \right)$$

This limiting frequency idea about probability is most often associated with the mathematician Richard von Mises (1883-1953).

An improper understanding of the law of averages can lead us into the *gambler's fallacy*. If a couple already has three boys and no girls and they are expecting another child, do you think it more likely that they will have a boy, that they will have a girl, or that they will be equally likely to have a boy or a girl? Before you answer, consider a dramatic evening at the casino in Monte Carlo on August 18, 1913. A roulette wheel when spun has equal chance of showing a red or a black number. On this evening, black came up 26 times in a row. What would you bet on as the most likely color for the next spin? A frantic rush to bet on red enriched the casino considerably; after all, a roulette wheel has no memory and will continue to produce black and red with equal probability in the future, regardless of its one-sided showing in the past. Although some medical evidence claims to support the theory of streaks of sexes in offspring of a family, this evidence is far from overwhelming; for the purpose of this example, we ignore such biological considerations and say that the parents of the three boys have the usual chance, $1/2$, to break their string.

Exercises

1.11   Out of 38 presidents of the United States, 13 were former vice presidents, In 1976 one television commentator estimated the odds that a vice president will become president to be 1:3, while a commentator for a different network estimated these odds to be 1:2. Which commentator was more nearly correct? (In terms of the classical definition, probability is the ratio of the number of favorable outcomes to the total number of possible outcomes, and odds are given as the ratio of the number of favorable outcomes to the number of unfavorable outcomes.) Does the classical definition of probability apply in this case? Does von Mises' definition apply?

1.12   Give an example of a real-world situation for which the classical definition of probability is not applicable. Is von Mises' definition applicable here?

1.13   Does the probability of rain as reported in the evening weather report illustrate either the classical or von Mises' definition of probability?

1.4   THE LANGUAGE OF PROBABILITY

The theory of probability as mathematical discipline should and can be axiomated in exactly the same sense as Geometry or Algebra.

Andreǐ Nikolayevich Kolmogorov

The advantage of the axiomatic method is that it allows the development of a general logical structure which can be applied to any specific situation which satisfies its assumptions. All results of plane geometry follow from Euclid's axioms regulating the elements *point* and *line*; now the entire apparatus can be applied to land surveying, navigation, building construction, crystallography, molecular structure, and on and on. Similarly, some questions about the representativeness of a jury may be answered by the same technique as questions about the toss of a coin.

In 1933 a Russian mathematician named Andreĭ Nikolayevich Kolmogorov (*b*. 1903) published the axiomatic structure of probability theory. A review of the essential concepts of set algebra is useful in understanding Kolmogorov's structure, since every uncertain situation involves a collection, a set, of possible outcomes. In this review, consider a class of 10 people with the following characteristics:

Example 1.4.1 Class data:

Initials	Sex	College	Year in School
M. B.	M	Engineering	Sophomore
S. R.	F	Liberal arts	Sophomore
G. S.	M	Engineering	Freshman
H. F.	M	Engineering	Junior
L. E.	F	Liberal arts	Junior
D. F.	F	Business	Sophomore
A. B.	M	Business	Freshman
P. L.	M	Engineering	Junior
B. H.	M	Engineering	Sophomore
T. M.	F	Liberal arts	Junior

A *set* is a precisely specified collection of objects, called *elements* of the set. In our example, elements are the members of the class. Other examples of elements of a set are

1. Television tubes coming off a production line
2. Residents of a community
3. The six faces of a die
4. All positive integers
5. All possible compositions of a town council for which an election will be held

We may specify a set by actually listing its elements or by describing them. In our example, the set may be written

{M.B., S.R., G.S., H.F., L.E., D.F., A.B., P.L., B.H., T.B.}

or equivalently

{people taking this class this semester}

A set B all the elements of which are elements of another set A is called a *subset* of A. We write B ⊂ A, read "B is contained in A," or A ⊃ B, read "A contains B." In our class, let F = {females} and L = {liberal arts majors}; then L ⊂ F.

In any discussion involving sets, there is some all-inclusive collection of objects to which the discussion is limited; this collection is called the *universe*, and we shall designate it by Ω. In our example, Ω might be all students at a university. For an opinion poll, Ω might be all registered voters. In probability theory, a common universe is the set of all logical possibilities of rolling two dice, where we may set Ω = {(1, 1), (1, 2), . . . , (1, 6), (2, 1), . . . , (6, 5), (6, 6)}.

For any subset A ⊂ Ω, the *complement* of A, written $\overline{A}$ (or not A), is the set of all elements of the universe that are not elements of the set A. In the class, $\overline{F} = M$ = {males}. The complement of Ω clearly contains no elements, and this will be called the null or empty set, written φ.

Often relationships among sets may be better understood by picturing them using *Venn diagrams*. The universe Ω is denoted by a rectangle and the sets under consideration by circles within the rectangle.

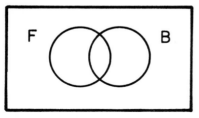

The relationship between the females, F, and the business majors, B, in our class can be drawn as shown here.

The two circles overlap, indicating that some, but not all, females are business majors and some, but not all, business majors are female.

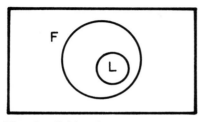

The relationship between F and L would be drawn as shown here.

That is, all liberal arts majors are female, but there are some females who major in other areas.

The set of all people who are *either* female *or* majoring in business is the *union* of F and B, written F ∪ B. This area is shaded in the following Venn diagram:

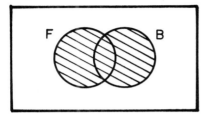

All female business majors (i.e., someone who is *both* female *and* majoring in business) make up F ∩ B, the *intersection* of F and B.

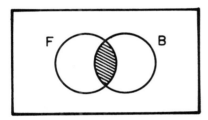

Two sets which have no elements in common are called *disjoint*. Here, F and A = {freshman} are disjoint; we would write F ∩ A = φ and draw the following diagram:

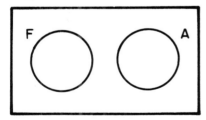

Consider the females in this class who are not in liberal arts:

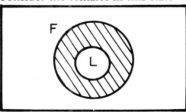

This set is written F − L and is the same as F ∩ L̄.

Now, how do we use all this language of sets to better our understanding of probability? To start with, probability people prefer their own vocabulary for these basic set notions:

Set vocabulary	Probability vocabulary
Universe	Sample space
Subset	Event
Disjoint	Mutually exclusive
Empty set	Impossible event

Also, in probability, the elements of the sample space are all possible outcomes of a random experiment, one whose outcome is subject to uncertainty. Let us now consider some examples.

Experiment	Sample space	Event
Roll two dice	$(1, 1), (1, 2), \ldots, (6, 6)$	"4" = $\{(1, 3), (2, 2), (3, 1)\}$
Select two people for a committee from a department of three men (labeled $1, 2, 3$) and two women $(4, 5)$	$(1, 2), (1, 3), (1, 4),$ $(1, 5), (2, 3), (2, 4),$ $(2, 5), (3, 4), (3, 5),$ $(4, 5)$	Both committee members male = $\{(1, 2), (1, 3), (2, 3)\}$
Toss a coin until a head appears	$\{TTT \ldots, H, TTH, \ldots\}$	Head appears on an even-numbered toss = $\{TH, TTTH, \ldots\}$
Study of the link between sex of student and undergraduate major	$\{$all college students$\}$	$\{$Females who major in math$\}$
Sex/major study described above	$\{$all degree programs$\}$	$\{$Major subject areas with male/female ratio different from that of the institution$\}$

The last two examples illustrate that in real-world problems, the definitions of sample space and event are not always uniquely determined. The description of the sample space should be appropriate to the experiment and to the context of the problem posed.

We now precisely define the concept of probability on the basis of our discussion of set relationships.

> **Probability: Axiomatic Definition** For any event A in a sample space Ω, the probability of A, denoted P(A), is a real number such that
>
> Axiom 1. $P(A) \geqslant 0$ for every event A in Ω.
> Axiom 2. $P(\Omega) = 1$.
> Axiom 3. If $A_1, A_2, A_3, \ldots$ are disjoint events, then
> $$P(A_1 \cup A_2 \cup A_3 \cup \cdots) = P(A_1) + P(A_2) + P(A_3) + \cdots.$$

This definition produces several useful properties of probabilities, most of which are strongly intuitive. We will illustrate by referring again to the class data of Example 1.4.1 where we will use the following set notation:

M = {males}, F = {females}
E = {engineering majors}, L = {liberal arts majors}, B = {business majors}
A = {freshmen}, B = {sophomores}, C = {juniors}, D = {seniors}

To compute the probability of selecting either a female or a freshman from the class, we may appeal to part 3 of the axiomatic definition, since F and A are disjoint (mutually exclusive) sets: $P(A \cup F) = P(A) + P(F) = 2/10 + 4/10 = 6/10$.

Notice that no seniors are taking this class; that is, $D = \phi$. So if we select one student at random, our chances of getting a senior are zero. We formalize this in the next result.

Theorem 1.4.1 The probability of the impossible event is zero; that is, $P(\phi) = 0$.

Proof. We can write $\Omega = \Omega \cup \phi$ (since ϕ adds no new elements to Ω) and $\phi = \Omega \cap \phi$ (since no elements are in both these sets). Axiom 3 implies that $1 = P(\Omega) = P(\Omega \cup \phi) = P(\Omega) + P(\phi)$, so $P(\phi) = 0$.

To calculate the chances of randomly selecting a male, we may count the males directly and obtain $P(M) = 6/10$, or equivalently we may count the number of females and reason that $P(M) = 1 - 4/10 = 1 - P(F) = 1 - P(\bar{M})$; that is, the probability of an event is 1 minus the probability of its opposite. We state this as the following theorem.

Theorem 1.4.2 The probability of the complement of any event A, denoted $\bar{A}$, is given by $P(\bar{A}) = 1 - P(A)$.

Proof. See Exercise 1.16.

Clearly, the chances we select any one from a subgroup of the sample space is a number less than one. (And yet we would wager that sometime one of you will someday compute a probability bigger than one and fail to recognize that your answer is unreasonable!) This is our next result.

Theorem 1.4.3   The probability of any event is a number between 0 and 1: $0 \leqslant P(A) \leqslant 1$ for every event A.

Proof.   See Exercise 1.17.

We see from the class data that all freshmen in the class are males, that is, $A \subset M$. Since we computed above that $P(M) = 6/10$, how would you estimate $P(A)$ relative to this figure? Clearly, $P(A) \leqslant 6/10$.

Theorem 1.4.4   If events A and B are such that $B \subset A$, that is, B is a subset of A, then $P(B) \leqslant P(A)$. (B, completely included in A, cannot have mass exceeding that of A, which contains it.)

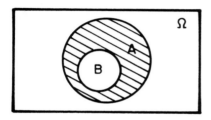

Proof.   The set A can be rewritten as the union of two disjoint sets: $A = B \cup (A \cap \bar{B})$. It therefore follows from Axiom 3 that

$$P(A) = P(B) + P(A \cap \bar{B})$$
or
$$P(B) = P(A) - P(A \cap \bar{B}) \qquad (1.4.1)$$

But $(A \cap \bar{B})$ is an event, and so $P(A \cap \bar{B}) \geqslant 0$, by Axiom 1. Thus $P(A) - P(A \cap \bar{B})$ has to be $P(A)$ or less. That is, $P(B) \leqslant P(A)$.

Now suppose you wanted to know your chances of selecting a female or a business major from this group; that is, $P(F \cup B)$. Is it appropriate to use Axiom 3 and calculate $P(F \cup B) = P(F) + P(B)$? Clearly not, since by doing so we are counting female business majors $(F \cap B)$ twice. We could correct this by subtracting them out once, and this is the point of our next result.

Theorem 1.4.5   For any two events A and B,

$$P(A \cup B) = P(A) + P(B) - P(A \cap B) \qquad (1.4.2)$$

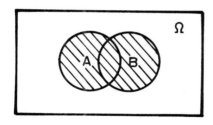

The points of the intersection A ∩ B are in both A and B, and thus P(A ∩ B) is part of both P(A) and P(B). Hence P(A) + P(B) includes P(A ∩ B) twice. Since we want probability mass counted only once in the union A ∪ B, we have to adjust P(A) + P(B) by subtracting P(A ∩ B).

Analytic proof. The union A ∪ B can be expressed as the union of *disjoint* sets:

$$A \cup B = A \cup (B \cap \bar{A})$$

and hence, by Axiom 3,

$$P(A \cup B) = P(A) + P(B \cap \bar{A}) \tag{1.4.3}$$

We can write B as the union of disjoint sets

$$B = (B \cap A) \cup (B \cap \bar{A}),$$

and so, by Axiom 3, have

$$P(B) = P(B \cap A) + P(B \cap \bar{A})$$

whence

$$P(B \cap \bar{A}) = P(B) - P(A \cap B) \tag{1.4.4}$$

Substituting (1.4.4) into (1.4.3), we obtain

$$P(A \cup B) = P(A) + P(B) - P(A \cap B)$$

These results, along with appropriate Venn diagrams, are helpful in sorting out probabilities when various conditions overlap. Consider a group of 150 mathematics students among whom 50 are taking a statistics course, 30 are taking a probability course, and 90 are taking neither statistics or probability. How many of the students are taking *both* statistics and probability?

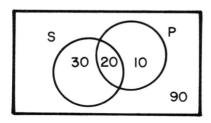

If 90 students are taking neither a statistics nor a probability course, then 150 – 90 = 60 students are taking one or the other (or both), so that the *union* of sets S (statistics) and P (probability) must contain 60 students. The Venn diagram shows how these must be arranged so that 50 are in S and 30 are in P. The result is that the intersection SP contains 20. Thus 20 students are taking both a statistics and a probability course.

The algebra of Theorem 1.4.5 will solve the problem formally since the *number of elements* in a set follows the same kind of logic as the *probability mass* in a set. If we use the notation N(A) to stand for the number of elements in A, then we have in the above situation $N(S \cup P) = 60$, $N(S) = 50$, $N(P) = 30$, and we argue

$$60 = N(S \cup P) = N(S) + N(P) - N(S \cap P)$$
$$= 50 + 30 - N(S \cap P)$$

whence

$$N(S \cap P) = 50 + 30 - 60 = 20$$

For an example with probability, take the case of a community where experience has shown that for families with school-age children the probability that the mother will attend a given PTA meeting is 0.80, the probability that the father will attend is 0.40, and the probability that both will attend is 0.30. What is the probability that at least one of the parents will attend the meeting?

If we designate the events (sets) as

M: mother attends meeting
F : father attends meeting

the diagram shows how the probabilities must be arranged, and hence shows that the probability is 0.90 that at least one of the two parents will attend, since that event is in the union M ∪ F. Algebraically we have

$$P(M \cup F) = P(M) + P(F) - P(M \cap F)$$
$$= 0.80 + 0.40 - 0.30 = 0.90$$

Exercises

1.14 A survey of 400 students reported that 200 smoked regularly, 150 drank coffee regularly, 90 both smoked and drank coffee regularly, and 180 neither smoked nor drank coffee. Are you suspicious of the results of this survey?

1.15 In a medical clinic 45 percent of all patients get shots and 20 percent of all patients get a shot and a drug prescription. Only 15 percent of all patients get neither a shot nor a drug. What proportion of patients get a drug prescription?

1.16 Provide a proof of Theorem 1.4.2. (Hint: Write Ω as the union of two disjoint sets.)

1.17 Prove Theorem 1.4.3. (Hint: Use Axiom 1 together with Theorem 1.4.2.)

1.18 Using Venn diagrams, illustrate that a distributive law holds for the intersection of an event A and the union of the two events B and C:

$$A \cap (B \cup C) = (A \cap B) \cup (A \cap C)$$

1.19 Extend the result of Theorem 1.4.5 to the union of three events by writing $A \cup B \cup C$ as the union of the two events A and $(B \cup C)$. Use Theorem 1.4.5 and Exercise 1.18 to arrive at the final result:

$$P(A \cup B \cup C) = P(A) + P(B) + P(C) - P(A \cap B) - P(A \cap C)$$
$$- P(B \cap C) + P(A \cap B \cap C)$$

1.20 Among the 120-person office staff of a company, 48 can type, 45 can take shorthand dictation, 40 can program a minicomputer, 12 can both type and take shorthand, 10 can both type and program, and 8 can program and take shorthand. How many talented people can perform all three jobs?

1.21 You walk into a casino, having assessed the probabilities that you will participate in the various indoor sports there:

Roulette only	0.60
Blackjack only	0.40
Slot machines only	0.50
Roulette and blackjack only	0.20
Roulette and slot machines only	0.30
Blackjack and slot machines only	0.10
Roulette and blackjack and slot machines	0.05

What is the probability that you will gamble (play at least one of the three games)?

1.22 To knock out a two-engine combat airplane, it is essential either to (1) destroy both engines, A and B, or to (2) destroy the pilot's cockpit. Of course, (1) and (2) together will do the job very nicely also. Let $A \cap B$ be the event described in (1), C be the event described by (2), and K be the event "knock out the plane"; then $P(K) = P((A \cap B) \cup C)$. Show that we can write

$$P(K) = P(A \cup C) + P(B \cup C) - P(A \cup B \cup C)$$

1.23   A plant has just completed a batch of 2000 heat exchangers. Of these, 15 have excessive curvature of the cap, 50 have holes in their protective covers, and 5 are defective for both reasons. If a potential customer selects one exchanger at random for inspection, what is the probability that it will be free of defects?

1.24   A firm interviews a large number of applicants for positions open on its sales force. Data indicate that in response to a question on leisure-time activities, 37.5 percent said they played golf, 75 percent said they bowled, and 12.5 percent did not mention any sport at all; no sport other than golf or bowling was mentioned. What percentage of the applicants both golf and bowl?

1.25   A store stocks calculators with the following inventory: 40 with memory, 25 with square-root key, 20 with percent key, 8 with memory and a square-root key, 6 with a percent key and memory, 6 with a percent and a square-root key, and 3 with all three capabilities. How many calculators are in the store's inventory?

1.26   The probability that on the first visit a minicomputer salesperson sells a computer to a prospective customer is 0.4. If the salesperson fails to make a sale on the first visit, the probability that the sale will be made on the second visit is 0.65. The salesperson never visits a prospective customer more than twice. What is the probability that the salesperson will make a sale to a particular customer?

1.5 SUBJECTIVE PROBABILITY AND ODDS

> *Kissinger's concern about a Russian attack on China was expressed many times. I used to tease him about his use of percentages. He would say there was a 60 percent chance of a Soviet strike on China, for example, and I would say, "Why 60, Henry? Couldn't it be 65 percent or 58 percent?*
>
>   H. R. Haldeman

Probabilities are quantitative expressions of uncertainty about a person's knowledge of the occurrence of some event. From this subjective point of view, we emphasize the uncertainty of our knowledge rather than the uncertainty of the event's occurrence. Even if the event has occurred, if we do not know about it the event will still be uncertain in our view. On the other hand, uncertainty of knowledge and uncertainty of occurrence coincide when the event is of the kind for which there is an empirically verified pattern of chance (e.g., toss of a coin, game of chance, sex of an unborn child).

In the preface of his stimulating treatise on probability, Bruno de Finetti (1974) displays the statement: "PROBABILITY DOES NOT EXIST." To understand this surprising statement, consider a recent decision by the U.S. Federal Power Commission to transport liquefied natural gas by marine tanker from the Atlantic Ocean through New York Harbor to a terminal on Staten Island, to be piped to New Jersey customers and reloaded onto barges for shipment up the East River. A consultant, Dr. T. W. Horner, reported a 1/4000 probability for a catastrophic accident from tanker movement in the next 10 years, and a 1/400 probability for such an event caused by barge movement. How could the probability of a liquefied natural gas catastrophe be a long-term relative frequency? The situation is unique: technology is new and will be improved in later decades. Commenting on the attempt to view the probability of a unique event as a relative frequency in a sequence of different universes, de Finetti writes of our "reluctance to abandon the inveterate tendency of savages to objectivize and mythologize everything."

To understand subjective assessments of probability, let us consider two opponents betting on the occurrence of some event A. To fix ideas, let us agree that you are one of the opponents, and that you are taking the side that A will occur (or has occurred). You and your opponent agree that if A occurs you will receive a certain payoff amount, and if A fails to occur you will forfeit a certain payoff amount. The relative sizes of the two amounts are part of the agreement depending on your (and your opponent's) assessment of the likelihood that A will or will not occur. Since probabilities of any kind are usually associated with occurrences of events under specific circumstances, we refer to "our probability" of an event A under a given set of circumstances H, and write $P(A|H)$; for example, P(point 7|fair role of two fair dice), P(head|fair toss of unbiased coin), or P(rain tomorrow|meteorologic conditions of today). The following definition makes these ideas precise.

An *elementary gambling situation* (EGS) is an agreement in which you pay u_1 dollars if the event A does not occur and you receive u_2 dollars if A does occur. The amounts u_1 and u_2 are called the *stakes* in the gamble; the ratio $u_2 : u_1$ is called the *odds*; and the fraction that your stake bears to the total is called the *betting quotient*,

$$q = \frac{u_1}{u_1 + u_2}$$

Now of course you would prefer that u_2 be very large compared to u_1, which is the same as saying that your stake be as small a fraction of the total stakes as possible, i.e., small q. But you have to take into account the other gambler, who is worried about the future just as you are. And both of you must assess the likelihood that A will occur.

Specifically, take the situation H to be the fair roll of a single die, and suppose the event A is "ace (one dot) is the result." What odds $u_2 : u_1$ do you want? Well, *want* is not quite right, for we all could *want* odds like 1000:1; it's more a matter of what odds would you like to try to get? You certainly start by ruling out even odds, 1:1. That makes $u_1 = u_2$, meaning you would get the same amount of money for rolling an ace as you would have to pay out if ace did not occur; but the nonoccurrence is much more likely than occurrence— a bad bargain for you!

It's easier to argue in terms of the betting quotient q, the fraction of total stakes which you are willing to put up, because in *any* gamble that fraction $[u_1 /(u_1 + u_2)]$ is a number between 0 and 1. You would like q as close to zero as possible, and your opponent would like it as close to one as possible. Suppose, for instance, you put up \$1 (i.e., $u_1 = 1$) at a q value of 0.05. This makes

$$0.05 = q = \frac{1}{1 + u_2}$$

$$1 + u_2 = \frac{1}{0.05} = 20$$

$$u_2 = 20 - u_1 = 20 - 1 = 19 \qquad \text{[Notice that odds are 19:1.]}$$

So if the ace is rolled, you win \$19 while having to pay only \$1 if ace fails to show—a fine deal for you. Near the other extreme, a q value of 0.9 when $u_1 = 1$ gives

$$0.9 = q = \frac{1}{1 + u_2}$$

$$1 + u_2 = \frac{1}{0.9} = 1.11$$

$$u_2 = 1.11 - 1 = 0.11 \qquad \text{[Odds are 0.11:1.]}$$

If ace is rolled you will be paid 11¢, whereas if ace fails you shell out your \$1— a bad bargain.

It seems that your attitude toward the bet ranges from displeasure to pleasure as the value of q (the fraction of total stakes to be committed by you) decreases from one toward zero. We feel intuitively that there must be some value of q at which your displeasure:pleasure is at a kind of break-even point, a position where you would view your likely gain and likely loss as balancing out. This would be the place where you would be *indifferent* as to which side of the bet you'd take: bet on the occurrence of A at odds $u_2 : u_1$ or bet on the nonoccurrence of A at odds $u_1 : u_2$. At such a point you would see your status and your opponent's status as balanced. For such a value of q the bet is called a *fair bet*, and the stakes set up as $u_2 : u_1$ are called fair-betting odds.

The value of q for such a fair bet in an EGS model accounts for the relative likelihoods of the occurrence and nonoccurrence of the event A. Take the case of rolling one die with A being the event "the roll is ace." Assuming the die to be perfectly balanced and the rolling of it to be an honest, unbiased roll, our previous intuition tells us to expect that the die face with one dot, one of the six faces of the die, will show up approximately once in six tosses, while some other face will show up approximately five times in six tosses. In the example of q = 0.05 considered earlier in our EGS discussion, we saw that if you would pay $1 for a failure you would receive $19 for a success. Thus just one toss of ace would make up for 19 tosses of non-ace, so that in 20 tosses you would on the average need just one ace in order to break even. But in 20 tosses you can expect something over three aces (one per six tosses, approximately). This means you expect to average $57 winnings in 20 tosses while paying out $17 (17 failures at $1 each)—a really good deal. In the example of q = 0.9 we saw that if failure costs $1, winning would give you 11¢. Here one failure wipes out nine successes. So in 10 tosses you need 9 successes to break even. But with the die-roll bet you see a future of something less than 2 successes in the 10 rolls (1 in 6). Hence you're facing 22¢ winnings and $8 losses in 10 rolls—a very undesirable deal.

Intuitive ideas of fairness (and self-protection) soon suggest that if each toss has one outcome (ace) in your favor and five outcomes (2, 3, 4, 5, or 6 dots) in favor of your opponent, then you should put up 1/6 of the stakes and your opponent should put up 5/6. We can check out the reasonableness: $u_1 = 1$ and $u_2 = 5$; you anticipate on the average one ace in six tosses, and so you anticipate that on the average six tosses will result in $5 winnings (one ace at $5 payoff) and $5 losses (five nonaces at $1 payout each). You see that your opponent has the same balance: for an average of six tosses he or she wins five times, getting from you $1 each time, and loses once, paying you $5 that time. Thus you would consider it fair to bet on the occurrence of ace at odds 5:1, or on the nonoccurrence of ace at odds 1:5, and you could be indifferent as to which side you take in the bet.

Now the basis of your judgment of fair is your assessment of the *probability* of A, $P(A)$, and if you are satisfied that it is fair for you to put up the fraction q of total stakes (and for your opponent to put up the fraction $1 - q$), then you are taking q as *your* value of $P(A)$. That is, out of N repeatable bets on A we expect to win the fraction $P(A)$ of them and to lose $1 - P(A)$. Thus we expect to win $N[P(A)]$ times, receiving u_2 each time and to lose $N[1 - P(A)]$ times, each time paying out u_1. So if winnings are to equal losses (whereby the bet is fair) we have to anticipate

$$N[P(A)]u_2 = N[1 - P(A)]u_1$$

and this gives

$$[P(A)]\, u_2 \;=\; [1 - P(A)]\, u_1 \;=\; u_1 - [P(A)]\, u_1$$

so that

$$u_1 \;=\; [P(A)]\, u_2 + [P(A)]\, u_1 \;=\; [P(A)]\, (u_1 + u_2)$$

or

$$P(A) \;=\; \frac{u_1}{u_1 + u_2} \;=\; q$$

Thus we have the following definition of probability.

Definition   Your probability $P(A)$ of the event A is the fraction q of total stakes you are willing to wager on the occurrence of A in what you regard as a fair bet in an EGS.

A very useful variation of this definition is phrased in terms of the betting odds $u_2 : u_1$.

Definition   If you consider $u_2 : u_1$ fair betting odds for betting on the occurrence of A in an EGS, then your probability $P(A)$ of the event A is

$$P(A) \;=\; \frac{u_1}{u_1 + u_2} \tag{1.5.1}$$

The relationship between odds and probability can be useful in two directions. Suppose we are invited to bet on an event A in a case in which the odds in favor of A are 3:2. We might agree with our opponent that our winning in case of success and our payout in case of loss be, respectively, $3 and $2, or $37.50 and $25, or any other amounts in the ratio 3:2. In the case of a $25 bet, our probability of winning is $P(A) = 2/(2 + 3) = 2/5$. This is a fair bet since if we wager N times we anticipate winnings of

$$\left(\frac{2}{5}\right)(N)(\$37.50) \;=\; N(\$15)$$

and losses of

$$\left(\frac{3}{5}\right)(N)(\$25) \;=\; N(\$15)$$

Proceeding in the reverse direction, suppose we believe that A has a 20 percent chance of occurring; that is we begin with our assessment of $P(A)$ as 0.2. What odds should we regard as fair? Since

$$P(A) = \frac{u_1}{(u_1 + u_2)}$$

we have

$$u_1 [P(A)] + u_2 [P(A)] = u_1$$

or

$$u_2 = \frac{u_1 [1 - P(A)]}{P(A)}$$

From this, we have the odds

$$\frac{u_2}{u_1} = \frac{1 - P(A)}{P(A)} = \frac{P(\bar{A})}{P(A)} \qquad (1.5.2)$$

In the case of $P(A) = 0.2$, we should demand odds of

$$\frac{1 - 0.2}{0.2} = \frac{0.8}{0.2} = \frac{4}{1}$$

Every experienced gambler knows: fair odds are the ratio of failure probability to success probability.

Once you have determined your judgment of fair odds, you are in a position to bet with increased vigor if you are offered better odds, or to retreat from the bet if the odds offered are worse. Better and worse are easily identified: remembering that the odds fraction u_2/u_1 is the payoff ratio, any fraction larger than the fair one is better than fair for you, any fraction smaller than the fair one is worse. Here is a case where bigger is better. Take the example of odds 3:2. The payoff fraction is $3/2$, so that if you wager $10 you will receive $3/2$ times $10, that is $15, if you win. For any fraction larger than $3/2$ a win will pay you more than $15 and for any fraction smaller than $3/2$ a win will pay you less than $15. So if you have judged $15 fair, you can tell what better and worse are.

A few more words about gambling terminology are in order. We have referred to the odds ratio $u_2 : u_1$ as the ratio of your opponent's stake to your stake, but of more direct interest to you has been the ratio of your winnings if A occurs to your loss if A does not occur. The reciprocal of the odds ratio, $u_1 : u_2$, could be used as well; the important thing to keep straight is what each ratio implies about your winnings and losses. Almost invariably a gambling casino, race track, or sports-event betting pool will quote odds in the order $u_2 : u_1$. This is the reason for using that convention. The casino, track, or betting pool is in clear reality your opponent and quotes the odds in terms of your potential payoff. Since your probability of A and the odds for a fair bet on A are related by the equality

$$\frac{u_2}{u_1} = \frac{1 - P(A)}{P(A)} = \frac{P(\overline{A})}{P(A)}$$

the reciprocal $u_1 : u_2 = P(A)/P(\overline{A})$ is called the *odds in favor of A*. If you are contemplating betting on A, either odds will give you the information you need, but you must be sure about which you agree to. In standard EGS betting you bet on A at stated *odds against A*. For example, if my fair odds (against A) are $2/3$, then my $P(A) = 3/(2 + 3) = 3/5$. And, the other way around,

$$\frac{1 - P(A)}{P(A)} = \frac{1 - 3/5}{3/5} = \frac{2/5}{3/5} = \frac{2}{3}$$

which represents the odds. If my odds in favor of $\overline{A}$ are $1/6$ (small), then my $P(A) = 6/7$ (large); if the odds are $6/1$ (large), then my $P(A) = 1/7$ (small). We recognize our intuitive logic: if odds against us are large, our probability of winning is small, and vice versa.

Let us now summarize your situation for the general case where you have the *assets* (or *capital*) Y to bet on A. If the odds are $u_2 : u_1$, you will win $Y(u_2/u_1)$ if A occurs, and lose Y if $\overline{A}$ (i.e., not A) occurs. This is the *change in your assets*, with winnings a positive value and loss a negative value. If you lose, then you have $Y - Y = 0$ as net total assets. If you win you have total assets amounting to

$$Y + Y\left(\frac{u_2}{u_1}\right) = Y\left(1 + \frac{u_2}{u_1}\right) = Y\left(\frac{u_1 + u_2}{u_1}\right)$$

If you have considered $u_2 : u_1$ fair odds, then $(u_1 + u_2)/u_1$, in the last equation, is the reciprocal of *your* probability $P(A)$ of A.

So we can now collect the essential facts about your assets, chances, and probability as follows.

If you have assets in the amount Y to bet on the occurrence of event A and the odds against A are $u_2 : u_1$, then

$$\text{Change in assets} = \begin{cases} Y\left(\dfrac{u_2}{u_1}\right) & \text{if A occurs} \\ -Y & \text{if } \overline{A} \text{ occurs} \end{cases} \quad (1.5.3)$$

$$\text{Total assets} = \begin{cases} Y\left(\dfrac{u_1 + u_2}{u_1}\right) & \text{if A occurs} \\ 0 & \text{if } \overline{A} \text{ occurs} \end{cases} \quad (1.5.4)$$

If and only if the odds $u_2 : u_1$ are *fair* odds against A, then your probability of A, $P(A)$, enters the situation in which the following results are equivalent to those above:

$$\frac{u_1}{u_1 + u_2} = P(A) \qquad (1.5.5)$$

$$\frac{u_2}{u_1} = \frac{P(\bar{A})}{P(A)} \qquad (1.5.6)$$

$$\text{Change in assets} = \begin{cases} Y \dfrac{P(\bar{A})}{P(A)} & \text{if A occurs} \\ -Y & \text{if } \bar{A} \text{ occurs} \end{cases} \qquad (1.5.7)$$

$$\text{Total assets} = \begin{cases} \dfrac{Y}{P(A)} & \text{if A occurs} \\ 0 & \text{if } \bar{A} \text{ occurs} \end{cases} \qquad (1.5.8)$$

Let us consider some examples.

Example 1.5.1   We consider it a fair bet to bet \$16 on event A provided if A occurs we are paid \$4 (and our \$16 is returned) and if A does not occur we lose the \$16. What is our probability of the event A?

Solution.   Here \$16 = Y, our assets, \$4 is the change in assets if A occurs, and \$20 is our total assets if A occurs. Hence, by Equation (1.5.8) we have $16/P(A) = 20$, or $P(A) = 16/20 = 4/5$.

Example 1.5.2   At what odds (against A) are we playing in Example 1.5.1?

Solution.   Once we know $P(A) = 4/5$, Equation (1.5.6) gives us the odds against A:

$$\frac{u_2}{u_1} = \frac{P(\bar{A})}{P(A)} = \frac{1 - 4/5}{4/5} = \frac{1/5}{4/5} = \frac{1}{4}$$

or 1:4. Note how these low odds go along with high probability (4/5).

Example 1.5.3   Suppose I know that $P(A) = 2/3$. Should I be pleased or displeased to bet \$10 on A, provided I am paid \$17 if A occurs and lose my \$10 if A does not occur?

Solution.   Here $P(A) = 2/3$, Y = 10, total assets = 17 if A occurs. In case of a *fair* bet, Equation (1.5.8) gives

$$\text{Total assets} = \frac{10}{2/3} = 10\left(\frac{3}{2}\right) = 15 \quad \text{if A occurs}$$

Since I am offered 17 as total assets if A occurs, the bet is better than fair, and I should be pleased.

Example 1.5.4 Suppose I am offered the same terms as in Example 1.5.3 to bet on a different event B, whose likelihood I am sure is greater than A's. Should I be more pleased or less pleased with this bet than with the one on A?

Solution. Under terms of a fair bet my total assets if B occurs should be $Y/P(B) = 10/P(B)$. Being sure that $P(B)$ is greater than $P(A)$, I am sure that $P(B) > 2/3$ and so $10/P(B) < 10/(2/3)$. Hence, my total assets if B occurs would be *less than* \$15 under a fair bet. Thus the offered \$17 is even better for me than it was when I considered the A bet, which had a total asset potential of \$15 under fair betting. I should be more pleased with the bet on B than with the bet on A.

Example 1.5.5 A person is displeased with a gamble of betting \$40 on an event A when the terms specify that if A occurs he is to be paid \$80, and he is to lose his \$40 if A does not occur. Is the person's $P(A)$ equal to $1/2$, greater than $1/2$, or less than $1/2$?

Solution. Since the person is displeased with the terms, it must be that the specified total assets if A occurs are less than what the person would require in a fair bet. That is, according to Equation (1.5.8),

$$80 < \frac{Y}{P(A)} = \frac{40}{P(A)}$$

so that $P(A) < 1/2$.

   All our arguments involving assets have been based on the tacit assumption that the *utility* (value) of, say, 100 monetary units is 100 times the utility of 1 unit. In practice this is not always so and utility depends heavily on the individual and the circumstances. The cases illustrated in Figure 1.2 illustrate several possible modes of utility. Case a represents fair or linear utility; here the utility of any monetary amount is directly proportional to the amount. Case b represents a reckless gambler who tends to overestimate the positive utilities and underestimate the negative ones. Conversely, a cautious, conservative person who is opposed to gambling will underestimate the positive utilities and overestimate the negative ones (case c). For a poor person (case d) the values of both positive and negative utilities are overestimated, while a rich person (case e) can afford to underestimate both of them. It has been determined experimentally that there is a common attitude toward utilities represented by case f: each one of us is a gambler to a certain point, after which we

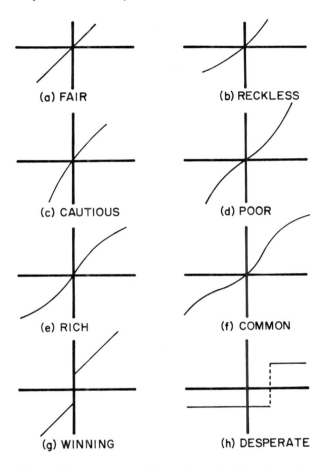

(a) FAIR

(b) RECKLESS

(c) CAUTIOUS

(d) POOR

(e) RICH

(f) COMMON

(g) WINNING

(h) DESPERATE

Figure 1.2 Some possible modes of utility dependent on the character of an individual. (Based on Kemeny and Thompson, 1957.)

tend to be overly cautious. The threshold point depends, of course on the individual. Next there is a winning (case g) attitude toward utilities promoted by some sport coaches. (Winning is not everything, it is the only thing.) The mere fact of winning, no matter how small the amount, yields a certain fixed positive utility, and the mere fact of losing even a negligible amount results in a definite negative utility. Finally we have case h of a desperate attitude when a certain fixed amount of money is urgently required: no lesser amount will do, while a larger amount is inconsequential for this particular instance. The reader is invited to devise additional realistic modes of utility.

The subjective probability in terms of fair betting odds which we have been discussing in this section is entirely consistent with the axiomatic structure of Section 1.4. The event A on which you consider betting, its complementary failure $\overline{A}$, and any other events entering into the EGS together make up a set of events satisfying Axiom 1. The procedure of taking our probability P(A) as the fraction of total stakes which we are willing to wager on A effectively satisfies Axioms 2 and 3. Subjective probabilities are not arbitrary assessments but are based on rational behavior.

Exercises

1.27 Fair odds in favor of an event A are 1:2. What is the probability of A? How would the odds be quoted in the usual manner of odds *against* A? If the odds against A were 1:2, what would be the probability of A?

1.28 Much to the distress of many people, in August 1978 the British book-making firm Ladbroke's was taking bets on who would succeed the recently deceased Pope Paul VI. Ladbroke's listed Italian Cardinal Sergio Pignedoli as the 5:2 favorite, "Which means," as an Associated Press bulletin explained, "that a $2 bet would win $5 if Pignedoli were selected." What was Ladbroke's probability of Pignedoli's selection?

1.29 One casino puts the odds in favor of an event A at 2:3, while another casino puts them at 5:9. In which casino should you bet?

1.30 Nevada roulette consists of spinning a wheel which has 38 positions of equal size marked on its circumference. One position is marked "0" and another "00"; these two positions are colored green. The remaining 36 positions are marked 1, 2, 3, . . . , 36, arranged irregularly around the circumference, with position colors alternating red and black. Show, using the classical definition of probability under assumption of perfect symmetry and balance of the wheel, that the probability of a marker's stopping at position "0," say P(0), is 1/38, and of its stopping on a red position, say P(red), is 9/19. From this find the odds for a fair bet for you to place on red. The actual house odds in a casino are 1:1; are these favorable or unfavorable to you?

1.31 The probability of an event A is 0.3. You are offered a bet such that if A occurs your net profit is $10, while if A does not occur your net loss is $5. Would you consider this a fair bet? (*Net profit* and *net loss* are terms often applied respectively to *positive change in assets* and the *magnitude of negative change in assets*.)

1.32 George Jones considers it a fair bet to bet $10 on an event A provided he is paid $5 and his wager is returned if A occurs, and he loses his $10 if A does not occur. What is Jones' probability of the event A? Will Jones be willing or unwilling to bet $25 and receive a total of $35 if A occurs?

1.33   Should you be pleased to bet $20 on event A provided you are paid the *total* of $80 if A occurs and lose your $20 if A does not occur, given that your assessment of likelihood is P(A) = 1/5?

1.34   Smith is pleased to bet $5 on event C under the conditions that she is paid $10 and the bet is returned if C occurs, and she loses her $5 if C does not occur. Is Smith's assessment of P(C): P(C) < 1/3, P(C) = 1/3, or P(C) > 1/3?

1.35   The odds in favor of event B are 6½:5½. If the event occurs, you win $55. How much should your fair bet be?

1.36   The probability of an event A is 0.4. You are offered a bet whereby your net profit is $10 if A occurs and your net loss is $8 if A does not occur. What odds (against A) are you being offered? Are you willing to make the bet?

1.37   Ms. Doe is displeased by the prospect of betting $7 on the event D under terms which would pay her $21 and return her bet if D occurs, while taking away her $7 as loss if D fails to occur. Is Doe's assessment of P(D): P(D) < 1/4, P(D) = 1/4, or P(D) > 1/4?

1.38   The announced odds on an event (i.e., the odds against the event) are 4:3. If the event occurs, your net win will be $20. What must be the size of your bet?

1.39   Before the division playoffs in a recent year the following were the odds against the winning of the World Series by each one of the stated teams: Phillies 5:1, Dodgers 7:5, Yankees 8:5, Royals 2:1.
   a.   Compute the probability of winning the World Series for each of the teams, assuming the odds fair. (Keep in mind that since the odds were quoted before division playoffs, the resulting probabilities are not a collection that adds up to one.)
   b.   Suppose you had bet $10 on the Phillies, $15 on the Dodgers, $20 on the Yankees, and $5 on the Royals, and suppose furthermore that the Yankees won the Series. How much money would you have won (or lost) as a result of these four bets?

1.40   *Newsweek* (November 16, 1981) reported on the odds that London bookmakers offered on the sex of the baby then expected by Prince Charles and Lady Diana—10:11 odds on the baby being a boy, even bets on a girl, and 50:1 on twins. If you had placed a $2 (£1) bet on each possibility, calculate your net winnings (losses), since Prince William was born on June 21, 1982.

2

REASSESSING PROBABILITIES

Probability is founded on the presumption of a resemblance between those objects of which we have had experience and those of which we have had none.

David Hume

2.1 CONDITIONAL PROBABILITY AND INDEPENDENCE

Now that we are able to assess probabilities, let us consider revising our probabilities in the light of additional information we might obtain about the uncertain situation. In fact, all probabilities are conditional on the current state of information. Returning to the class data of Example 1.4.1, let us ask whether there might be a relationship between a person's sex and choice of major in school. For simplicity, we classify both males and females as either liberal arts (L) or other ($\bar{L}$) majors. We may organize the data into a table (Table 2.1) to reflect these two characteristics of interest. That is, three of our class are female liberal arts majors, six are males with other majors, and so on.

Suppose we choose a student at random from the class. The classical definition of probability suggests that the probability this person is a liberal arts major is $P(L) = 3/10 = 0.3$; that is, 30 percent *of the class* are liberal arts majors.

Table 2.1

	Females	Males	Both
L	3	0	3
$\bar{L}$	1	6	7
Both	4	6	10

30

Now, suppose we ask what fraction *of the females* major in liberal arts. This is clearly 3 out of 4 or 0.75; that is, if we are selecting at random from the females, our chances of selecting a liberal arts major are 75 percent rather than 30 percent from the entire class. This probability from the subset 0.75, is the *conditional probability* of L given that F has occurred and we write P(L|F) = 0.75. There are six males, none of whom are liberal arts majors, so that P(L|M) = 0/6 = 0. Thus we may speak of the probability of selecting a liberal arts major in three different contexts depending on the additional information we are given.

Liberal arts majors

Of the class	Of the females	Of the males
P(L) = 0.3	P(L|F) = 0.75	P(L|M) = 0

Our chances of selecting a liberal arts major apparently differ depending on which group we use as a basis for our selection.

Considering P(L|F) again, we may write:

$$P(L|F) = \frac{3}{4} = \frac{3/10}{4/10} = \frac{P(L \cap F)}{P(F)}$$

and this result becomes our definition of conditional probability.

> **Conditional Probability**   For any two events A and B, the conditional probability of A given that B has occurred is written P(A|B) and is defined as
>
> $$P(A|B) = \frac{P(A \cap B)}{P(B)}$$
>
> provided that P(B) > 0.

Example 2.1.1   Weather station records in a certain region give the probability of a clear day in January as 0.60 and the probability of a warm, clear day in January as 0.15. You look out your window one January morning and see that the day is clear. What are the chances that the day will be warm? Let us write W as the event "the day is warm" and C as "the day is clear." We have been given by the weather station the following information:

$$P(W \cap C) = 0.15$$
$$P(C) = 0.60$$

and we are asked to find the probability of the event "clear day will be warm," which is P(W|C). According to the definition of conditional probability,

$$P(W|C) = \frac{P(W \cap C)}{P(C)} = \frac{0.15}{0.60} = 0.25$$

Let us organize the data from this exercise in a table:

	W	$\overline{W}$	Both
C	0.15		0.60
$\overline{C}$			
Both			1.00

Since the entries within the body of the table are probabilities that two events occur simultaneously or jointly (e.g., 0.15 is the probability that a day is jointly clear and warm), we call this a *joint probability table*. The probability that a clear day will be warm is the proportion of warm days in the C row of this table. A very different conditional probability would be the probability that a warm day will be clear; that is, before you open your curtains one January morning, you hear a warm temperature reported on the radio. What are the chances such a day will be clear? To answer this question we would need to obtain from the weather bureau the probability of a warm day in January, P(W), and enter this value in our table. Suppose we find that P(W) = 0.20. The table can now be completely filled in by subtraction:

	W	$\overline{W}$	Both
C	0.15	0.45	0.60
$\overline{C}$	0.05	0.35	0.40
Both	0.20	0.80	1.00

The numbers in the margins of the table, as 0.60 = P(C) and 0.80 = P($\overline{W}$), are called *marginal probabilities*. The conditional probability that a warm day will be clear, P(C|W) can now be found:

$$P(C|W) = \frac{P(C \cap W)}{P(W)} = \frac{0.15}{0.20} = 0.75$$

Notice the difference between the value of P(C|W) and that of P(W|C) computed earlier. Conditional probabilities must be interpreted carefully.

Example 2.1.2   Recalling our old example of rolling a single, balanced die, consider the outcomes in which the face showing is even, E, and in which the face showing is a prime number, B. We then have the following probabilities:

Outcome of a roll	Event	Probability
Even number	$E = \{2, 4, 6\}$	$P(E) = 3/6$
Prime number	$B = \{1, 2, 3, 5\}$	$P(B) = 4/6$
Even prime number	$E \cap B = \{2\}$	$P(E \cap B) = 1/6$

A friend rolls the die and tells you that the outcome is a prime number; what are the chances that the outcome is even? This is asking for the probability of an even outcome, given that the outcome is prime:

$$P(E|B) = \frac{P(E \cap B)}{P(B)} = \frac{1/6}{4/6} = \frac{1}{4}$$

What about the different event that an even outcome is a prime number? This is

$$P(B|E) = \frac{P(B \cap E)}{P(E)} = \frac{1/6}{3/6} = \frac{1}{3}$$

If we rearrange the statement of the definition of conditional probability, we obtain another useful rule of probability:

> **General Multiplication Rule of Probability** For any two events A and B with $P(B) > 0$,
>
> $$P(A \cap B) = P(A|B)P(B)$$

This result, sometimes called the law of compound probability, is rather more intuitive than the previous definition from which it comes. If 55 percent of a population are male and 10 percent of all males are color-blind, then what percentage of the population are color-blind males? If you answered 10 percent of 55 percent, or 5.5 percent, then you were making use of the multiplication rule of probability. Let M = the event "male" and C = the event "color-blind"; then the question asks for $P(C \cap M)$, which by the result above is $P(C \cap M) = P(C|M)P(M) = (0.10)(0.55) = 0.055$. Pick an adult at random and you have a 5.5 percent chance of having a color-blind male.

It is not uncommon in our consideration of probabilities for information about a single event A to come to us in terms of its joint relationship with another event B. The following Venn diagram shows that we can write A as

$$A = (A \cap B) \cup (A \cap \bar{B})$$

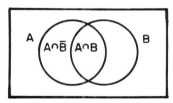

These two components of A are mutually exclusive, so that by the definition of probability (Axiom 3) we may write

$$P(A) = P(A \cap B) + P(A \cap \bar{B}) \qquad (2.1.2)$$

The multiplication rule allows us to rewrite Equation (2.1.2):

$$P(A) = P(B)\,P(A|B) + P(\bar{B})\,P(A|\bar{B}) \qquad (2.1.3)$$

Equation (2.1.3) is called the *law of inflating probabilities* by many modern writers.

Example 2.1.3 The probability of your friend's plane arriving at the airport on time is 80 percent if there is no precipitation, 20 percent if there is precipitation. You listen to a weather report and hear that there is a 40 percent chance of precipitation. What do you consider to be the probability that your friend's plane will arrive on time? Let A be the event that the plane arrives on time and R be the event of precipitation. Then, from Equation (2.1.3),

$$\begin{aligned}
P(A) = P(A \cap R) + P(A \cap \bar{R}) &= P(R)\,P(A|R) + P(\bar{R})\,P(A|\bar{R}) \\
&= (0.40)(0.20) + (0.60)(0.80) \\
&= 0.08 + 0.48 = 0.56
\end{aligned}$$

There is a 56 percent chance that the plane will arrive on time.

The concept of conditional probability allows us to formulate models for practical problems in which results at some stage may be influenced by previous occurrences. We consider an example of a class of problems involving the idea of a directed graph.

Example 2.1.4 [Based on a discussion by Ortel and Rossi (1981).] A mouse runs a maze between feeding stations to entertain people touring a laboratory. The following information is posted:

> The mouse always begins feeding at station a, after which he is equally likely to go to station b or c. If he feeds at b, his next stop is station e_1 with probability 1/3 or to station d with probability 2/3. If he feeds at c, he then goes to e_4 with probability 1/3 or to e_5 with probability 2/3. If he feeds at d, then he is equally likely to go to e_2 or e_3. The mouse ends the performance at any of the stations e. In return for a donation you will receive a 3-year subscription to our journal if the mouse runs the path $abde_2$.

This information can be summarized in the figure shown here, called a *directed tree*.

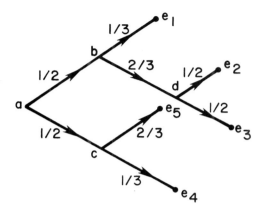

We would like to estimate the number of subscriptions that the lab can expect to give away; that is, what is $P(abde_2)$? Let a corresponding upper-case letter denote the event of all complete paths through a fixed small letter; that is, $B = \{abe_1, abde_2, abde_3\}$. The information stated in the problem can now be written in terms of conditional probabilities: $P(A) = 1, P(B|A) = P(C|A) = 1/2$, $P(D|B) = P(E_5|C) = 2/3, P(E_2|D) = P(E_3|D) = 1/2, P(E_1|B) = P(E_4|C) = 1/3$. Now the path $abde_2$ can be written in terms of the event $A \cap B \cap D \cap E_2$, and the general multiple rule for these four events gives

$$P(abde_2) = P(A)\, P(B|A)\, P(D|A \cap B)\, P(E_2|A \cap B \cap D)$$

Since $A \cap B = B$ and $A \cap B \cap D = D$, we have that

$$P(abde_2) = (1)\,(1/2)\,(2/3)\,(1/2) = 1/6$$

Roughly $1/6$ of the donors will receive complimentary subscriptions.

In applying probability theory to a practical situation, it is obviously important to clarify the relationship between the abstractions of the theory and the applied statement of the problem. Exercise 2.20 asks you to investigate the implications of this statement when the mouse runs a slightly altered maze.

Recall again our class of Example 1.4.1, where the probability of selecting a liberal arts major varied depending on additional information we were given about the sex of the group from whom we selected. It seems that finding a liberal arts student is *dependent* on the sex of the people from whom we select. Had this not been the case (that is, in a situation where sex and major are *independent*), our chances of finding a liberal arts major should be about the same whether we select from the entire class or from a subset of a single sex. Thus, A is independent of B if the conditional probability of A (given B) is the same as the unconditional probability of A: $P(A|B) = P(A)$. Similarly,

B is independent of A if $P(B|A) = P(B)$. A consequence of this relationship is that if A is independent of B, then

$$P(B|A) = \frac{P(B \cap A)}{P(A)} = P(A|B) \frac{P(B)}{P(A)} = P(A) \frac{P(B)}{P(A)} = P(B)$$

That is, if A is independent of B, then B is independent of A.

> **Independent Events**   Two events A and B are independent if $P(A|B) = P(A)$.

Let us look again at the multiplication rule in light of our discussion of independence of events. Recall that the multiplication rule gives $P(A \cap B) = P(A) P(B|A)$. If, in addition, A and B are independent, then $P(B|A) = P(B)$, so that we have a multiplication rule for independent events.

> **Multiplication Rule for Independent Events**   If A and B are *independent* events, then
>
> $$P(A \cap B) = P(A) P(B)$$

Example 2.1.5   In a sequence of two fair tosses of a fair coin, are the events "head on the first toss" and "head on second toss" independent? The sample space of the sequence of tosses is composed of the four points HH, HT, TH, TT, and each point is assigned probability mass 1/4 under the assumption of fair coin and fair toss. Let events be designated

A = head on first toss
B = head on second toss

Then

$$P(A) = P(\{HH, HT\}) = 2\left(\frac{1}{4}\right) = \frac{1}{2}$$

$$P(B) = P(\{HH, TH\}) = 2\left(\frac{1}{4}\right) = \frac{1}{2}$$

$$P(A \cap B) = P(\{HH\}) = \frac{1}{4}$$

Now,

$$P(B|A) = \frac{P(B \cap A)}{P(A)} = \frac{1/4}{1/2} = \frac{1}{2} = P(B)$$

so that A and B are independent. We also see that $P(A \cap B) = 1/4 = P(A) P(B)$.

It is important to emphasize that the multiplication of unconditional probabilities is appropriate *only* for *independent* events; multiplication of probabilities for dependent events can lead to dangerous conclusions. *People v. Collins* (Fairley and Mosteller, 1974) is an actual case in which a major issue was one of probability. Eyewitnesses testified that a robbery in California had been committed by a black man with a beard and a blonde girl with a ponytail driving a yellow car. The defendants answered this description, but there was no other evidence against them. The prosecution assumed the following chances:

P(yellow car) = 0.1	P(girl with blond hair) = 0.33
P(man with beard) = 0.25	P(black man with beard) = 0.1
P(girl with ponytail) = 0.1	P(interracial couple in car) = 0.0001

When multiplied together, these chances amount to about 1 in 12 million. The prosecution stated that this would be the probability that any other couple possessed these characteristics; since the chances were so small, this couple was obviously guilty. The jury convicted. On appeal the California Supreme Court reversed the verdict, essentially for the following two reasons. First, this is an improper multiplication of probabilities; some of the events are clearly dependent. Second, the court found the stated probabilities subjective and found no evidence to support them. The prosecutor gave us an example of what *not* to do with probability.

It is easy enough to extend the multiplication idea for independence from two events to three events. We see

$$P(A \cap B \cap C) = P(A)P(B)P(C)$$

as our obvious criterion. But if A, B, and C are independent, we would certainly want to imply that A and B are independent, A and C are independent, B and C are independent. Sad to say, the above equality will not guarantee such independence, as the following example shows.

Example 2.1.6 Consider the following space with probability masses as shown.

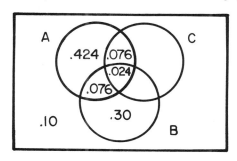

$$P(A) = 0.424 + 0.076 + 0.024 \\ + 0.076 = 0.60$$

$$P(B) = 0.30 + 0.076 + 0.024 \\ = 0.40$$

$$P(B) = 0.076 + 0.024 = 0.10$$

We have

$$P(A \cap B \cap C) = 0.024$$

while also

$$P(A)P(B)P(C) = (0.60)(0.40)(0.10) = 0.024$$

but

$$P(A \cap B) = 0.076 + 0.024 = 0.10 \quad \text{whereas} \quad P(A)P(B) = (0.60)(0.40) = 0.24$$

$$P(A \cap C) = 0.076 + 0.024 = 0.10 \quad \text{whereas} \quad P(A)P(C) = (0.60)(0.10) = 0.06$$

$$P(B \cap C) = 0.024 \qquad\qquad\qquad \text{whereas} \quad P(B)P(C) = (0.40)(0.10) = 0.04$$

Thus A, B, and C are *not* independent in pairs.

On the other hand, pair-wise independence does not guarantee that $P(A \cap B \cap C)$ will equal $P(A)P(B)P(C)$, as the following example illustrates.

Example 2.1.7 Consider the following probability space with probability masses as indicated.

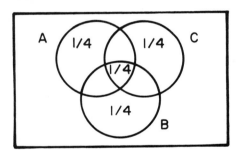

Here we have

$$P(A) = \frac{1}{4} + \frac{1}{4} = \frac{1}{2}$$

$$P(B) = \frac{1}{4} + \frac{1}{4} = \frac{1}{2}$$

$$P(C) = \frac{1}{4} + \frac{1}{4} = \frac{1}{2}$$

$$P(A \cap B) = \frac{1}{4} \qquad \text{while also} \qquad P(A)P(B) = \left(\frac{1}{2}\right)\left(\frac{1}{2}\right) = \frac{1}{4}$$

$$P(A \cap C) = \frac{1}{4} \qquad \text{while also} \qquad P(A)P(C) = \left(\frac{1}{2}\right)\left(\frac{1}{2}\right) = \frac{1}{4}$$

$$P(B \cap C) = \frac{1}{4} \qquad \text{while also} \qquad P(B)P(C) = \left(\frac{1}{2}\right)\left(\frac{1}{2}\right) = \frac{1}{4}$$

so that A and B are independent, A and C are independent, and B and C are independent. But

$$P(A \cap B \cap C) = \frac{1}{4} \quad \text{whereas} \quad P(A)P(B)P(C) = \left(\frac{1}{2}\right)\left(\frac{1}{2}\right)\left(\frac{1}{2}\right) = \frac{1}{8}$$

Thus, overall independence of more than two events requires the multiplication criterion to be satisfied for all compounds of two events, all compounds of three events, and so on.

Independence of Three or More Events The events $A_1, A_2, \ldots, A_k$ are independent if and only if the probability of the intersection of any combination of them is equal to the product of the probabilities of the single events.

Independence of events is a useful and powerful concept, significantly simplifying calculations involving complex probabilistic models. In fact, most real-world probabilistic models can be analyzed only if independence is assumed. For this reason, there is great temptation to abuse and misuse the concept. Whenever careful analysts make use of an assumption of independence, they go to great pains to ensure validity of the assumption. Numerous methods of scientific random sampling illustrate this concern.

When we are tossing a coin successively and observe after each toss whether it falls heads or tails, we obtain a sequence of trials which are usually assumed and verified experimentally to be independent. Even in this simple model there can sometimes be doubt whether the assumption of independence is valid, in particular if we use the same coin and it is tossed always by the same person. Tables of random numbers, careful shuffling of cards, and accurate balance of a well-built roulette wheel are aids to independence of draws, deals, and spins.

Erroneous thinking about independence works both ways. While scientific investigations often assume independence unjustifiably, there are cases in everyday life where people try to ignore independence when it is actually proved to be present. Many of us are unable to apply the idea of independence to things like the color of a traffic signal at our various arrivals, the sex of successive babies in a family, or occasions of bad luck or good luck. From the frightened airline traveler who always took a dummy bomb on board as a form of insurance, reasoning that it would be virtually impossible for *two* passengers with bombs to be on the same aircraft, it is not a long conceptual way to people who believe in absolutely safe gambling systems, to gamblers in casinos who keep roulette records, and to lottery and bingo experts who on the back of their tickets publish statistics of underrepresented numbers. Martin Gardner (1982) notes that "most people find it difficult to believe that the probability of an independent event is not somehow influenced by its proximity to other independent events of the same sort."

This brings us to the notion of *rare events*, which also causes some misunderstanding and unjustified conclusions on occasion. If we toss an unbiased coin 10 times and obtain heads on each toss, either we doubt that the coin is unbiased, or if we are completely sure that the coin is fair, we are extremely surprised that we have obtained such a rare combination of outcomes. (We return to this notion of rare outcomes in Section 5.5.) Indeed, under the independence and unbiasedness assumptions, the probability of the outcome of 10 heads in 10 tosses is $(1/2)^{10} = 1/1{,}024$; that is, the odds are more than 1000:1 against the occurrence of such a sequence. But consider now the situation when we observe the following sequence of heads and tails in 10 tosses: HHTHHTTHTH. The occurrence of this sequence does not produce any particular surprise and does not tempt us to modify our belief about the unbiasedness of a coin. However, the probability of obtaining this particular sequence is also (under the assumption of the independence and unbiasedness) $(1/2)^{10} = 1/1{,}024$, and this sequence therefore is as "rare" as the one involving 10 consecutive heads. What makes the second sequence less striking is the fact that it is one of many sequences involving six heads and four tails. There are indeed 210 such distinct sequences, and so the probability of obtaining one of the sequences involving six heads and four tails is $210(1/1{,}024)$, which is slightly larger than 0.2, making the occurrence of such an event not at all unusual.

The rarity of an event often depends on the definition of the event in the context of interest. The royal flush in five-card draw poker is indeed very rare (Table B.2 in Appendix B gives the probability 4/2,598,960), and the five-card hand (2, 3 of spades, 5 of hearts, 7 of diamonds, 6 of clubs) is even more rare. The latter is one of 1,302,540 "bust" hands, so that P(bust hand) = 1,302,540 (1/2,598,960) = 0.501 (not rare). In poker, "bust hand" and "royal flush" are events of interest, rather than any specific array of cards.

A similar point is made by David Marks (1980):

First we notice and remember matches, especially *oddmatches* whenever they occur. (Because a psychic anecdote first requires a match, and, second an oddity between the match and our beliefs, we call these stories oddmatches. This is equivalent to the common expression "an unexplained coincidence.") Second, we do *not* notice nonmatches. Third, our failure to notice nonevents creates the short-run illusion that makes the oddmatch seem improbable. Fourth, we are poor at estimating combinations of events. Fifth, we overlook the principle of equivalent oddmatches, that one coincidence is as good as another as far as psychic theory is concerned.

We discuss this point further in Section 5.4.

Exercises

2.1   In a certain primary election 40 percent of all registered voters appeared at the polls and voted in the Democratic primary. If registration is 60 percent Democrat, what is the probability that a registered Democrat chosen at random was a primary voter?

2.2   In a simplified version of bridge, two players A and B each draw one card in turn from a deck, and compare results. B wins the round if he or she has drawn a spade and A has not; if both have drawn spades, the round is a tie; otherwise, B loses the round. Find the probability that

    a.   B wins.

    b.   The round is a tie.

    c.   B loses.

2.3   One of a pair of dice is loaded in such a way that the chance of a 1 turning up is 1/5, the other faces being equally likely among themselves. The other die of the pair is loaded so that the chance of a 6 turning up is 1/5, the other faces being equally likely among themselves.

    a.   To what does this loading increase the probability of throwing a 7 with the pair of dice?

    b.   Would you be willing or unwilling to bet \$2 that the 7 will appear with this pair of dice, provided that if this event occurs your total assets will be \$11?

2.4   A certain vaccination procedure administers a first shot of vaccine followed by a shot of a second vaccine *if* there is no detectable reaction to the first; if there is a detectable reaction to the first shot, no second shot is given. The probability of reaction to the first vaccine is 0.7, and, for persons unreactive to the first vaccine, the probability of reaction to the second is 0.9.

    a.   What is the probability that a randomly selected individual will react to neither vaccine?

    b.   What is the probability that a randomly selected individual will react to the second vaccine only?

2.5   In the fall of 1981, U.S. journalistic media reported heated debate on the sale of Airborne Warning and Control System (AWACS) planes to Saudi Arabia. President Ronald Reagan openly and vigorously lobbied for the sale. Finally, on October 28, 1981, the sale was approved by the U.S. Senate, with 52 senators voting in favor and 48 voting against. According to Senator Packwood, "With 90% of the votes, the President was 80% of the deciding factor" (*Washington Post*, October 30, 1981). Using Packwood's assertion and conditional probability concepts, estimate how many senators voted for the sale solely in deference to Presidential pressure, disregarding the merits of the case.

2.6   You visit a friend who has twin children, neither of whom you have seen. The first child who comes into the room is a boy. What is the probability that the second twin is also a boy? [Assume that b is the probability of two boys, g is the probability of two girls, and that (boy, girl) is as likely as (girl, boy).]

2.7. *Time* magazine had the following report in its March 5, 1979 issue: "Republican Alan Greenspan figured that Carter stands at best a 50% chance of renomination and, if he passes that hurdle, only a 60-40 shot at beating the Republican candidate. Mathematically, that would work out to a 30% chance of going all the way." Can we "mathematically" agree?

2.8   Prove that if events A and B are independent, then also

    a. $\bar{A}$ and B are independent.

    b. $\bar{A}$ and $\bar{B}$ are independent.

(Hint: In question b use the fact from Section 1.4 that the complement of the union of two sets is also the intersection of their complements.)

2.9 a. If A and B are independent events, prove that

$$P(A \cup B) = P(A) + P(\overline{A})P(B) = P(B) + P(\overline{B})P(A)$$

  b. Find an analogous formula for $P(A \cup B \cup C)$ when A, B, and C are independent events.

2.10 In the game of blackjack, a player is playing alone with the dealer and only four cards are dealt. An untied blackjack occurs when the player is dealt an ace with a face card or a ten (a blackjack) and then the dealer fails to get a blackjack. What is the probability of this event?

2.11 Uri Geller, the controversial psychic and mind reader, was able to tell in eight consecutive experiments what number was showing at the top of a single die which had been placed inside a metal box and shaken by the experimenter. What is the probability of such an occurrence if Uri is only guessing, the die is fair, and the successive tosses are independent?

2.12 A coin is unbalanced so that you are as likely to get two heads in two successive independent throws as you are to get a tail in one throw. For this coin, what is the probability of getting a head in a single throw? (Hint: Solve the equation $p^2 = 1 - p$.)

2.13 From a population of families, each having three children, a family is selected at random and the sex composition of its children observed. Show that the following two events are independent in the probability sense:
  M = family has children of both sexes
  L = family has at most one girl
[Hint: Designating boy by B and girl by G, set up a sample space of eight equiprobable points (BBB), (BBG), etc.]

2.14 After a typist typed a lengthy article, he checked the typing and found 40 typing errors. Independently, the author checked a copy of the typed article and spotted 30 errors, 24 of which were also noticed by the typist. Estimate the number of typing errors which both the typist and author missed.
[Hint: Let the unknown total number of errors be N. Then we estimate P(typist finds error) = 40/N, etc. Using the multiplication rule of probabilities for independent events, set up an equation whose solution is our estimate of N.]

2.15 X and Y repeatedly throw a fair die, in that order. The first to throw a 6 wins. What are the odds? [Hint: Using X to designate "X throws a 6," Y to designate "Y throws a 6," consider the event "X wins" as the union of disjoint events X, $\overline{X}\overline{Y}X$, $\overline{X}\overline{Y}\overline{X}\overline{Y}X$, ... The formula for sum of a convergent infinite geometric progression will eventually be useful. Use a similar process for P(Y wins). This solution requires knowledge of advanced high-school algebra.]

2.16 In a box are n capsules containing the numbers 1, 2, 3, . . . , n (one number to a capsule). X and Y draw at random in that order, and keep a capsule each. Show that the probability that Y's capsule contains a higher number than X's is $1/2$. [Hint: Consider disjoint events (X draws 1 and then Y higher), (X draws 2 and then Y higher), and so on. When necessary use the formula $1 + 2 + \cdots + k = k(k + 1)/2$.]

2.17 Peter and John, in order, draw at random and keep a card from a bridge deck of cards. John wins if he draws a higher ranking card of the same suit as Peter, or if he draws a spade after Peter has drawn some other suit. What is the probability that John wins? (Hint: At some point the result in Exercise 2.16 is useful.)

2.18 If A and B are the teams playing in a World Series (where the team which first wins four games wins the Series), and we assume successive games independent, with A having probability p of winning each time and B's probability being $q = 1 - p$, find the probability that
a. A wins the series in four games, in five games, in six games
b. the series ends in four games, in five games, in six games.
What are the probabilities in question b if A and B are of equal strength? (Hint: Using W for win and L for loss, consider the various disjoint winning-sequence events like LWWWW, WLWLWW, etc.)

2.19 Using the frequency interpretation of probability, what is the exact proportion of children taking three tests that will typically fail to score above the thirty-sixth percentile on at least one of the tests? (Hint: "None" is the complement of "at least one.") Would it be reasonable to require a score above the thirty-sixth percentile for just two of the three tests?

2.20 Let us alter the course of the mouse in the maze of Example 2.1.4 so that if he feeds at c he then goes to e_4 with probability $1/3$ or to d with probability $2/3$. (Station e_5 is completely eliminated.) How does this alteration of the maze change our formulation of the model? What additional information would you require from the lab in order to compute the probability of a free subscription with this new maze?

2.2 BAYES' RULE

In a court of law, if a defendant is found guilty by the jury, what is the probability that the defendant is actually innocent? If J denotes the event "defendant found guilty by the jury" and G denotes that event "defendant is in fact guilty," then the event described above is $\bar{G}|J$ (not guilty given convicted by the jury). Our legal system views the conviction of innocent persons $(J|\bar{G})$ as a serious error and operates in a context designed to minimize its probability. Bayes' rule, due to an English clergyman Thomas Bayes (1702-1761), gives us a mechanism for determining $P(\bar{G}|J)$ if we know $P(J|\bar{G})$ and some related information.

From Section 2.1, recall the definition of conditional probability for any two events A and B. As a direct consequence of this definition, we can write

$$P(A|B) = \frac{P(A \cap B)}{P(B)}$$

and

$$P(B|A) = \frac{P(A \cap B)}{P(A)} = \frac{P(B)P(A|B)}{P(A)} = \left(\frac{P(B)}{P(A)}\right) P(A|B) \qquad (2.2.1)$$

If B can be partitioned into k mutually exclusive events $B_1, B_2, \ldots, B_k$, then

$$A = (A \cap B_1) \cup (A \cap B_2) \cup \cdots \cup (A \cap B_k) \qquad (2.2.2)$$

and these components of A are mutually exclusive, since the B_i's are. Thus, by the third axiom of probability,

$$P(A) = P(A \cap B_1) + P(A \cap B_2) + \cdots + P(A \cap B_k) \qquad (2.2.3)$$

Using the law of compound probability, Equation (2.2.3) can be written as

$$P(A) = P(B_1)P(A|B_1) + P(B_2)P(A|B_2) + \cdots + P(B_k)P(A|B_k) \qquad (2.2.4)$$

Then any of the components B_i has a probability conditional on A, and this gives us Bayes' rule.

Bayes' Rule   If $B_1, B_2, \ldots, B_k$ are mutually exclusive events, and A is any event with $P(A) > 0$, then for any B_i

$$P(B_i|A) = \frac{P(B_i \cap A)}{P(A)} = \frac{P(B_i)P(A|B_i)}{P(A)}$$

$$= \frac{P(B_i)P(A|B_i)}{P(B_1)P(A|B_1) + P(B_2)P(A|B_2) + \cdots + P(B_k)P(A|B_k)}$$

The statement of Bayes' rule may seem rather bulky, but it has important implications in the study of uncertain events. When B_i may be thought of as antecedent to A in time, then there may be interpretation of B_i and A as cause and effect, respectively. Then Bayes' rule is sometimes interpreted as the probability that B_i was the cause of A. In such a context, the unconditional probabilities $P(B_i)$ are called *prior* probabilities and the conditional probabilities $P(B_i|A)$ are called *posterior* probabilities.

Example 2.2.1   Returning to the legal situation with which we began this section, suppose that due to the severity of the error, the court sets $P(J|\overline{G}) = 0.01$; that is, no more than 1 out of every 100 innocent people should be erroneously convicted, in the long run. Acquitting a guilty person does not have the same ethical implications, so let us take $P(\bar{J}|G) = 0.05$; that is, $P(J|G) = 0.95$, or 95 percent of all guilty people are convicted. From historical records we see that 90 percent of all people brought to trial have been guilty; that is, $P(G) = 0.9$. Now, suppose that the court has just convicted a defendant; Bayes' rule allows us to calculate a posterior probability that the defendant is innocent:

$$P(\bar{G}|J) = \frac{P(\bar{G})P(J|\bar{G})}{P(\bar{G})P(J|\bar{G}) + P(G)P(J|G)} = \frac{(0.1)(0.01)}{(0.1)(0.01) + (0.9)(0.95)}$$

$$= 0.0012$$

That is, only about 1 in 1,000 convicted people were unjustly convicted by the jury, so the principles of our legal system seem to be upheld. These calculations have been, of course, very dependent upon our assessment of the proportion of all defendants who are guilty. You are asked to investigate this relationship further in Exercise 2.35.

Example 2.2.2   A company owns drilling rights off the coast of Britain. From geological records they know that in this area high-quality oil is found 20 percent of the time, low-quality oil is found 30 percent of the time, and no oil is found 50 percent of the time. Before drilling they have the option of performing a seismic test to obtain more information on the presence of oil. In the past this test has given positive results on 80 percent of the sites where high-quality oil was subsequently found, on 20 percent of the sites containing low quality oil, and on 20 percent of the sites containing no oil. The company has the test performed and the results are positive. What are their chances of finding oil, given this additional information?

Frequently, it is helpful to organize such a list of probabilities into a table and then to apply Bayes' rule systematically. Let B_1 be the event that high-quality oil is found, B_2 be the event that low-quality oil is found, and B_3 be the event that no oil is found; let A denote a positive test result. We construct the following table.

Event	I $P(B_i)$	II $P(A|B_i)$	III $P(A \cap B_i)$	IV $P(B_i|A)$
Good oil (B_1)	0.20	0.80	0.16	0.50
Poor oil (B_2)	0.30	0.20	0.06	0.19
No oil (B_3)	0.50	0.20	0.10	0.31
	1.00		0.32	1.00

The probabilities in columns I and II are stated in the problem. The joint probabilities of Column III are obtained by multiplying the corresponding entries of columns I and II, according to the general multiplication rule of probability. Column IV is obtained through Bayes' rule, by dividing the corresponding entry from column III by the total of that column. Finally, the answer to the question posed, $P(\text{oil}|A)$ is as follows:

$$P(\text{oil}|A) = P(B_1|A) + P(B_2|A) = 1 - P(B_3|A) = 0.69$$

Thus, the positive test results cause the company to revise their probability of finding oil from 50 percent (0.20 + 0.30) to 69 percent.

Example 2.2.3 At a plant that produces widgets, three machines (B_1, B_2, B_3) feed a packaging line. Machine B_1 is responsible for 40 percent of the widgets produced, machine B_2 supplies 25 percent, and machine B_3 supplies 35 percent. From past records, the percentage of widgets which are defective is 2 percent from machine B_1, 4 percent from machine B_2, and 3 percent from machine B_3. If we select an item at random from the packaging line and find it to be defective, calculate the probability that that widget has come from each machine. Let D denote the event of finding a defective widget.

Event	$P(B_i)$	$P(D|B_i)$	$P(D \cap B_i)$	$P(B_i|D)$
Machine B_1	0.40	0.02	0.0080	0.281
Machine B_2	0.25	0.04	0.0100	0.351
Machine B_3	0.35	0.03	0.0105	0.368
	1.00		0.0285	1.000

That is, 28.1 percent of the defective widgets came from machine B_1, 35.1 percent came from machine B_2, and 36.8 percent came from machine B_3. This example illustrates the alternative description of Bayes' rule as *probability of causes*.

Example 2.2.4 [Based on a discussion by Robbins and Van Ryzin (1975.] A paternity suit involves a child with a blood type (call it T) different from his mother's and thus inherited from his father. The man alleged to be the father denies the charge and is required to take a blood test. If his blood type is something other than T, then he is clearly not the boy's father. However, if his blood type is T, the probability that he is the father is increased. The purpose of this example is to assess this increased probability. In the analysis we use the facts that the blood type T occurs in 10 percent of the population; prior to the blood test, the probability of this man being the child's father is 0.5. Let us define events

    A: the man has blood type T
    B: the man is the father of the child

The information given implies that the prior probability $P(B) = 0.5$. Notice also that $P(A|B) = 1$, since the man must have blood type T if the woman's allegation is true, while $P(A|\bar{B}) = 0.1$, the same probability as in the entire population. Now, Bayes' rule gives our desired posterior probability:

$$P(B|A) = \frac{P(B)P(A|B)}{P(B)P(A|B) + P(\bar{B})P(A|\bar{B})}$$

$$= \frac{(0.5)(1)}{(0.5)(1) + (0.5)(0.1)} = \frac{0.5}{0.55} = 0.91$$

quite an increase over the prior assessment of this probability!

It is not difficult to find examples of the dangerous consequences of confusing $P(A|B)$ with the different probability $P(B|A)$. In Washington, D.C., for example, the events "U.S. Senator" (A) and "male" (B) are such that $P(B|A)$ is very near one, while $P(A|B)$ is practically zero. Another important illustration of this possible confusion occurs frequently in disease diagnosis.

Example 2.2.5 Consider the problem of screening for cancer. Let A be the event that a person has the disease and B be the event that the test shows a positive result. The diagnostic test has 2 percent false positives and 1 percent false negatives. The patient and the physician are primarily interested in the answer to the question: "If the test results are positive, what are the chances that the patient actually has the disease?" We summarize the relevant information below.

Event	Prior probability	Conditional probability	Joint probability
Patient has disease, A	$P(A) = 0.003$	$P(B|A) = 0.99$	$P(B \cap A) = 0.00297$
Patient does *not* have disease, A	$P(\bar{A}) = 0.997$	$P(B|\bar{A}) = 0.02$	$P(B \cap \bar{A}) = 0.01994$
	1.000		$P(B) \quad = 0.02291$

Then we calculate the desired posterior probability as

$$P(A|B) = \frac{0.00297}{0.02291} = 0.130$$

There is only a 13 percent chance of actually having the disease when the test indicates that you do! This result leads us to question the wisdom of a large-scale screening program for diseases with low prevalence.

In discussions of political and social questions, emotions often run high, and it becomes easy to stress the particular conditional probability that supports one's own argument. For example, take a question of racial discrimination.

Example 2.2.6   In one year, 20 blacks and 80 nonblacks applied for admission to a small college. Of that group, 15 blacks and 35 nonblacks were admitted. Let A be the event that an applicant was admitted, and let B be the event that an applicant was black. One set of relevant probabilities is:

$$P(B|A) = \frac{15}{50} = 0.3$$

$$P(\bar{B}|A) = \frac{35}{50} = 0.7$$
                                                                                    (2.2.5)

Since 0.7 is much larger than 0.3, one might be led to use this as evidence in favor of discrimination. But consider another set of conditional probabilities:

$$P(A|B) = \frac{15}{20} = 0.75$$

$$P(A|\bar{B}) = \frac{35}{80} = 0.44$$
                                                                                    (2.2.6)

These probabilities leave a different impression, and the "correct" set is the one *appropriate* to describe the policy at issue.

The ideas of this section which involve reassessing probabilities on the basis of outcomes of experiments are related to statistical hypothesis testing. We explore the relationship further in Section 5.5. Bayes' rule is not the only reasonable way to update subjective probabilities. See the illuminating paper by Diaconis and Zabell (1982).

Exercises

2.21   Let A be the event that you have enough money in your bank account, and let B be the event that you postdate your check. If $P(A) = 0.8$, $P(B|A) = 0.1$, and $P(B|\bar{A}) = 1$, compute $P(A|B)$ and interpret the numerical result.

2.22   Let A represent the event that a person is an ex-convict, B the event that he is a rapist. The following probability assessments are given:

$$P(A) = 0.01$$
$$P(B) = 0.001$$
$$P(A|B) = 0.9 \qquad \text{whence} \qquad P(\bar{A}|B) = 0.1$$

Compute $P(B|A)$ and thus verify that the probability that a given ex-convict is a rapist is less than 0.1. Find $P(A|\bar{B})$.

2.23 Let D represent the event that a person is sick with disease D, and *pos* denote the event that a test for the disease is positive (*neg* for negative is the event complementary to pos). The test is such that it gives only 1 percent false positives and 2 percent false negatives; that is,

$$P(pos|\overline{D}) = 0.01, \quad \text{whence} \quad P(neg|\overline{D}) = 0.99$$
$$P(neg|D) = 0.02, \quad \text{whence} \quad P(pos|D) = 0.98$$

What is the probability that a person whose test is positive has the disease [i.e., what is $P(D|pos)$] if
a. The prevalence of the disease is 8 per 100 [$P(D) = 0.08$]?
b. The disease is rare, with prevalence 1 per 1,000 [$P(D) = 0.001$]?
Explain the difference between the results in questions a and b.

2.24 Let G denote the event that a person is a university graduate student, and let B denote the event that a person is bright. Show, by taking $P(G) = 0.001$ and choosing reasonable estimates of $P(B|\overline{G})$, that $P(G|B)$ may be close to zero in spite of the (somewhat disputed) fact that usually $P(B|G) = 1$.

2.25 Assess the probabilities $P(A|B)$ versus $P(B|A)$ in the following cases.
a. A = person is on welfare, B = person is poor
b. A = person is a movie star, B = person has been divorced at least once

2.26 Assume that the last juror to be selected for a panel to serve in the trial of a certain accused person will be 1 of 10 candidates. Of the 10, 6 are young persons and the other 4 are older people, this distribution being fairly representative of the proportions of young and older among jurors in the county generally. A special survey in the county has shown that 30 percent of jurors are sympathetic to a defendant accused of the particular type of offense at issue in the present trial, and it is known that 65 percent of sympathetic jurors are older people. Given that the final juror chosen is an older person, what is the probability that he or she is sympathetic to the accused?

2.27 One of two strings (A, B) of Christmas lights is chosen at random. All the lights in string A are good ($\overline{D}$) while 1/4 of the lights in string B are defective (D). Next we choose at random a light from the selected string. If it proves to be a good light, what are the odds in favor of our having chosen the perfect string?

2.28 In a certain city 40 percent of the voting-age people are Democrats, 50 percent are Republicans, and 10 percent are Independents. Records show that in a particular election 80 percent of the Democrats voted, 60 percent of the Republicans voted, and 30 percent of the Independents voted. If a voting-age person in the city is selected at random and it is learned that he or she did not vote in the election what is the probability that he or she is
a. An Independent?
b. A Republican?
c. A Democrat?

2.29   Consider the search for a new, producing oil well, where a gusher is the optimum find but a more modest oil deposit is highly acceptable. Past experience in the region under investigation indicates that drilling wells at random will produce a gusher 8 percent of the time and a moderate deposit 24 percent of the time, while the remaining 68 percent of borings are dry. Geologists can supply a seismic test to check out potential boring sites. The test gives a positive result in 85 percent of cases where a gusher is known to be present, in 60 percent of cases of modest deposit, and in 25 percent of cases of dry wells. If a seismic test is positive, what is the probability that a boring will tap an acceptable oil deposit?

2.30   This exercise is one case of a more general *capture-recapture* technique. The Department of Wildlife needs to estimate the number of deer within a wildlife preserve. Deer do not generally cooperate by standing still to be counted, so a more subtle technique has been developed. Suppose that the preserve is known to contain either 1, 2, 3, 4, or 5 deer with equal probability. We choose a deer at random, tag it, and let it go. The next day we observe a deer from this preserve and note that it is not tagged. What are the posterior probabilities that the preserve contains 1, 2, 3, 4, and 5 deer, respectively?

2.31   *Identification Evidence.* The knife used in a certain murder had on its handle a palm print similar to that of the accused defendant in the court trial of the case. Designate events at issue as
          H: a palm print similar to the defendant's is on the knife
          G: the defendant used the knife (he is guilty)
It is reasonable to assume that if the defendant is guilty, then it is certain that a palm print similar to his is on the knife, i.e., $P(H|G) = 1$. Since a palm print cannot be identified as definitively as fingerprints, it is possible that someone other than the defendant put the similar print on the knife, i.e., $P(H|\bar{G}) > 0$. The court is interested in assessing the likelihood that the defendant is guilty given that there is a palm print similar to his on the murder weapon. Thus $P(G|H)$ is the important probability at issue, varying according to various values of $P(G)$ and $P(H|\bar{G})$. Verify the entries in the following table.

Posterior Probabilities $P(G|H)$

	Prior probability $P(G)$					
$P(H|\bar{G})$	0.01	0.1	0.3	0.5	0.7	0.9
0.5	0.020	0.182	0.462	0.667	0.824	0.947
0.3	0.033	0.270	0.588	0.769	0.886	0.968
0.1	0.092	0.526	0.811	0.909	0.959	0.989
0.01	0.503	0.917	0.977	0.990	0.996	0.999
0.001	0.910	0.991	0.998	0.999	0.9996	0.9999

2.32   Members of the Senate are to be polled regarding their views on oil legislation. Forty percent of the senators are Democrats and 18 percent of these Democrats have opposed such oil legislation in the past. Twelve percent of the Senate is comprised of Republicans who have opposed such oil legislation in the past. A senator is polled and found to be in favor of the oil legislation. What is the probability that person is a Democrat? (Assume senators are either Democrats or Republicans).

2.33   *The Monty Hall Problem* (Selvin, 1975a,b). On the television show "Let's Make a Deal," one of three boxes (A, B, C) contains the keys to a new Lincoln Continental. The other two boxes are empty. The contestant chooses box B. Boxes A and C remain on the display table, and host Monty Hall suggests that the contestant surrender his box for $500. The contestant refuses, and Monty opens one of the remaining boxes (box A); the box does not contain the coveted keys. Now Hall offers the contestant $1,000 to surrender his box. The contestant again refuses, but instead he asks to *trade* his box B for the box C remaining on the table. Monty exclaims, "That's weird!" Show that the contestant knew what he was doing. [Hint: Let B be the event that the contestant chooses box B, K the event that the keys are in box B, A the event that Monty opens box A after the contestant has chosen box B. It must be assumed that Monty knows where the keys are. Use Bayes' rule to calculate the posterior probability $P(K|B \cap A)$.]

2.34   *Extension of Monty Hall Problem.* Show that is there are *four* boxes in the above contest and Monty opens *two* of the three remaining after the contestant has chosen his box (say box B again under terms as above), then the contestant has probability 3/4 of winning if he can switch boxes at the end.

2.35   In what sense are the calculations of Example 2.2.1 dependent on our assessment of P(G), the proportion of all people tried who are guilty? Let P(G) = p, and compute $P(\overline{G}|J)$ as a function of p. Graph this function and interpret your results in the context of our legal system.

2.3   THE ART OF ASKING QUESTIONS

One of the most common means of gathering information is to ask people questions: surveys, polls, and exams are examples of this. The fact that people may not always answer questions truthfully, either by design or through mis-understanding, often contaminates results and clouds analysis of the information. Our understanding of probability provides a way to reduce some of this contamination.

Randomized Response

When we ask a sensitive question which might subject someone to embarrass-ment, social stigma, liability to arrest and prosecution, or the like, it is unreasonable to expect truthful answers. Self-protection is more compelling than honesty when a person is asked, "Do you smoke marijuana at least once a week?" or, "Have you had an abortion within the last year?" or, "Have you ever cheated on your income tax return?" or, "Did you receive unauthorized help on your last take-home examination?"

In many such examples, the objective is to estimate the proportion p of a population that belongs in the yes-response category. We do this by selecting a random sample of n people from that population and estimating p by the proportion of affirmative replies in the sample, $\hat{p}$:

$$\hat{p} = \frac{\text{number of yes replies}}{n}$$

But if we cannot depend on getting a truthful response, the estimate is obviously inaccurate (biased). The technique of *randomized response* provides a way for a respondent to be completely honest while also being completely safe from identification.

To illustrate, let us pose two questions: (1) the sensitive question, say S, and (2) an innocuous question, say T, whose probability of yes-response in the population is known (θ). For example, we might pose:

  S: Do you smoke marijuana?
  T: Does your social security number end in an even digit? ($\theta = 1/2$)

The respondent is told to select randomly one of the questions, S or T, and answer truthfully. The questioner knows the answer, but not to which question it applies. For example, the questioner might instruct the respondent: "Toss a coin, being sure that only you see the result. If the coin lands heads, answer question S. If the coin lands tails, answer question T. I will hear your answer but will not know which question you are answering."

Let us use the following notation:

p_s = P(yes|S) = proportion of population having sensitive characteristic

θ = P(yes|T) = proportion of population having innocuous characteristic

π = P(S) = probability that random device indicates respondent should answer question S

λ = P(yes) = probability that respondent gives a yes-reply

Then

λ = P(respondent answers yes to S *or* answers yes to T)

 = P(S and yes) + P(T and yes)

 = P(S)P(yes|S) + P(T)P(yes|T)

 = $\pi p_s + (1 - \pi)\theta$ (2.3.1)

Solving this equation for the unknown p_s, we obtain

$$\pi p_s = \lambda - (1 - \pi)\theta$$

$$p_s = \frac{\lambda - (1 - \pi)\theta}{\pi} \tag{2.3.2}$$

Here π and θ are known, and we estimate λ by use of the proportion of yes replies in the sample, say

$$\hat{\lambda} = \frac{\text{no. of yes replies in sample}}{\text{no. of persons in sample}} \tag{2.3.3}$$

Then we have an *estimate* $(\hat{p}_s)$ of p_s:

$$\hat{p}_s = \frac{\hat{\lambda} - (1 - \pi)\theta}{\pi} \tag{2.3.4}$$

Example 2.3.1

> S: Do you smoke pot?
> T: Does your telephone number end in an even digit?

The randomizing device was an opaque plastic box containing 35 red and 15 white balls (all of the same small size and weight) and a small window against which a single ball could be seen after the box was shaken. The instructions were: If red ball appears, answer question S; if white ball appears, answer question T. The sample was a group of 100 students in an eastern university. The number of yes responses was 44.

$$\pi = P(\text{red ball}) = \frac{35}{50} = 0.7$$

$$\theta = P(\text{even terminal digit in telephone number}) = 0.5$$

$$\hat{\lambda} = \frac{44}{100} = 0.44$$

$$\hat{p}_s = \frac{0.44 - (0.3)(0.5)}{0.7} = \frac{0.44 - 0.15}{0.7} = \frac{0.29}{0.7} = 0.414 = 41.4\%$$

That is, we estimate that 41.4 percent of the population smokes marijuana—but we do not know which 41.4 percent of our sample does! A recent study of drug use among Canadian high-school students (Goodstadt, Cook, and Gruson, 1978) showed that the randomized response method indeed does encourage coopera- tion: an increase was shown in both the response rate of the students and in the mean reported drug use for this method over direct questioning.

One problem with the use of this model is that even with randomization, for many characteristics of interest (e.g., cheating, abortions) only a yes response could possibly subject the person to embarrassment or stigma. Let us consider replacing the irrelevant question T by the negation of the sensitive question S. For example,

S: Have you received unauthorized help on any exam this term?
T: Have you adhered to the honor code on every exam this term?

In this case, the relation between λ and p_s is obtained by replacing θ by $1 - p_s$ in Equation (2.3.1):

$$\lambda = \pi p_s + (1 - \pi)(1 - p_s) \tag{2.3.5}$$

If $\pi \neq 1/2$ (e.g., we do not flip a coin to decide which question to answer), we can estimate p_s by using $\hat{\lambda}$ of Equation (2.3.3) to estimate λ. If $\pi = 1/2$, then $\lambda = 1/2$ for any value of p_s, and so p_s cannot be estimated.

The method of randomized response can be generalized, by using two independent samples, to the case in which the probability θ of yes response to the innocuous question T is unknown, and to the case in which only one question, the sensitive one, is asked but confidentiality is still assured (see Warner, 1965; Folsom et al., 1973). The details of the practical aspects and the reliability of randomized response are investigated in an enlightening paper by Brewer (1981) discussing results of a survey of marijuana usage in Australia using randomized techniques as well as conventional survey methodology. One of the paradoxical findings was the reluctance of nonusers to admit the truth!

Multiple-Choice Questions

Most of us have had first-hand experience with examinations containing battalions of questions, each with four or five optional answers, of which customarily only one is the correct answer. Test analysts have determined that of all types of short-answer questions, the multiple-choice type with five optional answers is the most valid to test an individual's true knowledge. Test analysts, instructors, and students are quite interested in sorting out the likelihoods of real knowledge and of guesswork in such examinations. Let us consider a simplified version of this problem to avoid getting lost in complex details.

Assume that we have a multiple-choice exam consisting of two questions, each having three stated answer options, only one of which is the correct answer. We are interested in estimating the proportion of students who actually know the answers to both questions, basing that estimate on the distribution of submitted responses. We assume that a person who knows the correct answer will choose and report the correct option, and that a person who does not know the correct answer will guess correctly with probability $1/3$ and incorrectly with probability $2/3$.

Probabilities relative to a respondent drawn at random from the population of examinees are as follows:

p_{cc} = P(responses are: correct, correct)

p_{cw} = P(responses are: correct, wrong)

p_{wc} = P(responses are: wrong, correct)

p_{ww} = P(responses are: wrong, wrong)

Behind these probabilities are the probabilities relative to the actual knowledge of the random respondent:

λ_{11} = P(correct answer known for both questions)

λ_{10} = P(correct answer known for question 1 but not question 2)

λ_{01} = P(correct answer known for question 2 but not question 1)

λ_{00} = P(correct answer known for neither question)

From the formula for total probability, we can express each p in terms of the λ's:

$$p_{cc} = \lambda_{11}(1)(1) + \lambda_{10}(1)\left(\frac{1}{3}\right) + \lambda_{01}\left(\frac{1}{3}\right)(1) + \lambda_{00}\left(\frac{1}{3}\right)\left(\frac{1}{3}\right)$$

$$p_{cw} = \lambda_{11}(1)(0) + \lambda_{10}(1)\left(\frac{2}{3}\right) + \lambda_{01}\left(\frac{1}{3}\right)(0) + \lambda_{00}\left(\frac{1}{3}\right)\left(\frac{2}{3}\right)$$

$$p_{wc} = \lambda_{11}(0)(1) + \lambda_{10}(0)\left(\frac{1}{3}\right) + \lambda_{01}\left(\frac{2}{3}\right)(1) + \lambda_{00}\left(\frac{2}{3}\right)\left(\frac{1}{3}\right)$$

$$p_{ww} = \lambda_{11}(0)(0) + \lambda_{10}(0)\left(\frac{2}{3}\right) + \lambda_{01}\left(\frac{2}{3}\right)(0) + \lambda_{00}\left(\frac{2}{3}\right)\left(\frac{2}{3}\right)$$

that is,

$$p_{cc} = \lambda_{11} + \frac{1}{3}\lambda_{10} + \frac{1}{3}\lambda_{01} + \frac{1}{9}\lambda_{00}$$

$$p_{cw} = \frac{2}{3}\lambda_{10} + \frac{2}{9}\lambda_{00}$$

$$p_{wc} = \frac{2}{3}\lambda_{01} + \frac{2}{9}\lambda_{00} \qquad\qquad (2.3.6)$$

$$p_{ww} = \frac{4}{9}\lambda_{00}$$

The examination results give us estimates of the p's: the proportion of respondents whose answers to the two questions are correct is an estimate of

p_{cc}, say $\hat{p}_{cc}$; similarly for observed and calculated $\hat{p}_{cw}$, $\hat{p}_{wc}$, $\hat{p}_{ww}$. When these estimates are used to replace the p's in Equation (2.3.6), we then can put a hat (^) on each λ and solve the system of equations for the $\hat{\lambda}$'s. Notice that the last equation gives $\hat{\lambda}_{00}$ immediately, and then we move rapidly up the array of equations by successive substitution and solution. In this way we estimate the unknown (and unobservable) λ's.

Example 2.3.2   In an experiment on such a two-question examination, 27 answer cards were distributed as follows:

Question 1	Question 2	No. of respondents
Correct	Correct	20
Correct	Wrong	3
Wrong	Correct	3
Wrong	Wrong	1

The Equations (2.3.6)—hatted—are then

$$\hat{p}_{cc} = \frac{20}{27} = \hat{\lambda}_{11} + \frac{1}{3}\hat{\lambda}_{10} + \frac{1}{3}\hat{\lambda}_{01} + \frac{1}{9}\hat{\lambda}_{00}$$

$$\hat{p}_{cw} = \frac{1}{9} = \frac{2}{3}\hat{\lambda}_{10} + \frac{2}{9}\hat{\lambda}_{00}$$

$$\hat{p}_{wc} = \frac{1}{9} = \frac{2}{3}\hat{\lambda}_{01} + \frac{2}{9}\hat{\lambda}_{00}$$

$$\hat{p}_{ww} = \frac{1}{27} = \frac{4}{9}\hat{\lambda}_{00}$$

The last equation gives immediately

$$\hat{\lambda}_{00} = \left(\frac{1}{27}\right)\left(\frac{9}{4}\right) = \frac{1}{12}$$

Then in the third equation we obtain

$$\frac{1}{9} = \frac{2}{3}\hat{\lambda}_{01} + \left(\frac{2}{9}\right)\left(\frac{1}{12}\right)$$

$$\frac{1}{9} = \frac{2}{3}\hat{\lambda}_{01} + \frac{1}{54}$$

$$6 = 36\hat{\lambda}_{01} + 1$$

$$\hat{\lambda}_{01} = \frac{5}{36}$$

and, since the second equation, for $\hat{\lambda}_{10}$, is the same as the equation for $\hat{\lambda}_{01}$, we have

$$\hat{\lambda}_{10} = \frac{5}{36}$$

Then the first equation yields:

$$\frac{20}{27} = \hat{\lambda}_{11} + \left(\frac{1}{3}\right)\left(\frac{5}{36}\right) + \left(\frac{1}{3}\right)\left(\frac{5}{36}\right) + \left(\frac{1}{9}\right)\left(\frac{1}{12}\right)$$

$$\frac{20}{27} = \hat{\lambda}_{11} + \frac{5}{108} + \frac{5}{108} + \frac{1}{108}$$

$$80 = 108\hat{\lambda}_{11} + 5 + 5 + 1$$

$$\hat{\lambda}_{11} = \frac{69}{108} = \frac{23}{36}$$

It is interesting to note that although $\hat{p}_{cc} = 20/27 = 74.1\%$, the estimate is $\hat{\lambda}_{11} = 23/36 = 63.9\%$; in this case the test seems to be overestimating the knowledge of the examinees.

Misclassification Probability

The last result emphasizes that in any examination where answers are objective in nature, there is nonzero probability that a correct answer will be given by guessing rather than by actual knowledge. Thus there is the likelihood that an examinee can be misclassified as to his or her knowledge. Probability analysis can indicate the potential magnitude of such likelihood.

Consider first an examination of five multiple-choice questions, each question having four optional answers, only one of which is correct. The scoring corresponds to the number of questions answered correctly; for example, a respondent who answers three questions correctly gets a score 3. What is the probability that an examinee who *knows* correct answers to only two questions gets a score of 3?

The examinee scores 2 by the answers to the two questions for which he or she has knowledge, and scores 1 by guessing correctly the answer to one of the three questions for which he or she does not have knowledge. Thus, if we designate an examinee's score by S and the number of questions for which the correct answer is known by r, then the required probability is

$$P(S = 3 \mid r = 2) = P(1 \text{ out of 3 replies is guessed correctly})$$

The probability of choosing "at random" the one correct option out of four is $p = 1/4$ for any given question. If then we designate a correct choice by c and an incorrect choice by w, the probability of "1 out of 3 correct" is the probability that the examinee's sequence of answers to the three questions at issue is cww, wcw, or wwc, and each of these sequences has probability $(1/4)^1 (3/4)^2$ under assumption of independent guessing on each question. Hence,

$$P(S = 3 | r = 2) = P(\{cww, wcw, wwc\} | p = 1/4)$$

$$= 3\left(\frac{1}{4}\right)\left(\frac{3}{4}\right)^2 = \frac{27}{64} = 0.422$$

As another example, what is the probability that the examinee considered above would score 4? In this case two correct answers come by knowledge and the remaining two correct responses come by guessing two out of the three questions on which knowledge is lacking.

$$P(S = 4 | r = 2) = P(2 \text{ out of 3 replies are guessed correctly})$$

$$= P(\{ccw, cwc, wcc\} | p = 1/4)$$

$$= 3\left(\frac{1}{4}\right)^2 \left(\frac{3}{4}\right)^1 = \frac{9}{64} = 0.141$$

Finally, for such an examinee,

$$P(S = 5 | r = 2) = P(\{ccc\} | p = 1/4) = \left(\frac{1}{4}\right)^3 = \frac{1}{64} = 0.016$$

$$P(S = 2 | r = 2) = P(\{www\} | p = 1/4) = \left(\frac{3}{4}\right)^3 = \frac{27}{64} = 0.422$$

Table 2.2 collects such results for all possible values of r and S. The reader is urged to verify the entries. From this table we see, for example, that on the average only 42 percent of examinees who really know answers to just two questions will get the correct grade score 2, while 58 percent will receive an unjustified higher grade.

Table 2.2

S	r					
	0	1	2	3	4	5
0	0.237	0	0	0	0	0
1	0.396	0.316	0	0	0	0
2	0.264	0.422	0.422	0	0	0
3	0.088	0.211	0.422	0.562	0	0
4	0.015	0.047	0.141	0.375	0.750	0
5	0.001	0.004	0.016	0.063	0.250	1

Exercises

2.36   The sensitive question in a randomized response survey is: "Have you had an abortion?" The randomizing device is a box containing 10 red balls, 15 white balls, and 20 blue balls. The instructions are: "Draw a ball at random from the box, making sure that no one but you sees its color. If the ball is red, mark the answer yes. If the ball is white, mark the answer no. If the ball is blue, truthfully answer the stated question." In one sample of 125 women there were 60 yes responses. Estimate the population proportion p_s of abortions in the population.

2.37   If there are r possible answers to a multiple-choice question and λ is the proportion of the population who know the correct answer, show that the proportion p_c who answer correctly is given by $p_c = \lambda + (1 - \lambda)/r$.

2.38   Suppose that in a single multiple-choice question with three options offered for the answer, the three choices A, B, and C were selected with equal frequencies (1/3) by the examinees. Show that this indicates (provided that the sample of examinees is large) that at most a negligible portion of the subjects knew the correct answer.

2.39   Suppose that in another single question with three multiple-choice options for the answer, answer A is selected 90 percent of the time, answers B and C each 5 percent of the time. Show that this indicates that 85 percent of the examinees knew the correct answer and only 5 percent guessed it correctly without knowing. (Assume the correct answer is A.)

2.40   Each of 150 customers was presented with two glasses of syrup, one containing real maple and the other containing artificial flavoring, and was asked to discriminate between the two syrups. The experiment was conducted twice, and 30 customers discriminated wrongly in both trials. Estimate the number of people among the 150 customers who can truly discriminate between the two syrups. Based on this survey, would you recommend to the company that it replace the maple in the syrup by artificial flavoring? (Hint: Use the situation of Exercise 2.37 with r = 2.)

2.41   In a multiple-choice examination consisting of 10 questions, each having five optional answers of which one and only one is the correct answer, what is the probability that an examinee who knows the correct answer for six questions will score
a.   8 correct?
b.   9 correct?

3

CONTROLLING UNCERTAINTY

3.1 EXPECTATION OF RANDOM VARIABLES

If you had it to do over, would you change anything? "Yes, I wish I had played the black instead of the red at Cannes and Monte Carlo."

Winston Churchill

In earlier chapters we have assessed and reassessed probabilities of outcomes in uncertain situations. Some of these situations have resulted in outcomes that can be described numerically: the number of spots on the face of a die, the points scored by a winning baseball team. In other situations, the outcomes themselves may not be numeric, but we may be interested in a numeric characteristic of a particular outcome: the percentage of women on the city council, the average starting salary after graduation of students in a particular degree program.

> A function that assigns a number to each outcome of a random experiment is called a *random variable*.

Example 3.1.1 A 50-year-old man purchases a life insurance policy. Whether the policy is collected in the subsequent year is an uncertain situation which depends upon whether the man lives or not, a random event. The insurance company's profit, let's call it X, for that customer in that year is a number whose value depends upon the outcome of that random event; hence it is a random variable. From life tables we find the probability that a 50-year-old man lives another year (and thus does not collect on his policy) to be 0.995. Let us simplify the insurance company's rates somewhat and suppose that they charge each 50-year-old male $520 a year for a $100,000 life insurance policy. Now

Table 3.1 Distribution of Insurance Company Profit

Outcome	Value of X	Probability
Man lives	$520	0.995
Man dies	$520–$100,000	0.005

we can make a table of possible values of company profit X and the probability associated with each value (Table 3.1). The two right columns of Table 3.1 form the probability distribution of X.

> The *probability distribution*, or merely *distribution*, of the random variable X is given by a list of the values of the random variable with corresponding probabilities which sum to 1.

Returning to our insurance company, out of all of its 50-year-old male policy holders, it could "expect" to make $520 on 0.995 (99.5%) of them and $-$99,480 (pay out this net amount) on 0.005 (0.5%) of them. That is, if the company sells such policies to N such men it could expect a profit per person of

$$\frac{(\$520)(0.995)N + (-\$99,480)(0.005)N}{N}$$

$$= (\$520)(0.995) + (-\$99,480)(0.005) = \$20$$

Almost without realizing it, we have formulated the definition of expectation of a random variable.

> If X is a random variable which can assume the values x_1, x_2, x_3, ..., x_k with respective probabilities $P(x_1)$, $P(x_2)$, ..., $P(x_k)$, such that $P(x_1) + P(x_2) + P(x_3) + \cdots + P(x_k) = 1$, then the *expectation* (or *expected value*) of X, designated $E(X)$, is
>
> $$E(X) = x_1 P(x_1) + x_2 P(x_2) + x_3 P(x_3) + \cdots + x_k P(x_k)$$

The concept of expectation is perhaps older than that of probability (Daston, 1980). In 1657 Christian Huygens (1629-1695), a Dutch mathematician, published the first definition of expectation: "One's Hazard or Expectation to gain any Thing is worth so much as, if he had it, he could purchase the like Hazard or Expectation again in a just and equal Game." From this circular definition the concept of expectation evolved to denote a set of rules for determining the fairness of an agreement involving uncertainty or for assessing the advisability of participating in a risky venture (e.g., in games of chance or the sale of an insurance policy). The probabilistic definition of expectation given above arose from the application of expectation to contracts, depending on uncertain outcomes, among which games of chance were included. Let us consider a specific example.

Example 3.1.2 We propose to you a new game. You roll two dice; if the point (sum of the numbers showing) is either 6, 7, or 8, we win. If it is 2, 3, 4, 5, 9, 10, 11, 12, you win. Since you have lots more winning combinations than we do, we propose that you pay us $2 when we win and we pay you $1 when you win. Do you want to play?

In order that you will not be led to think all random variables must be called X, let us have W denote our winnings for this game. We first construct the probability distribution of W. One possible value for W is $2, and the probability associated with this value is the probability that the point rolled is 6, 7, or 8. In Section 1.2 we listed all 36 possible points in the throw of two dice, and we recall part of that list here:

Point	Outcome	Probability
6	$(1, 5)$, $(5, 1)$, $(2, 4)$, $(4, 2)$, $(3, 3)$	5/36
7	$(1, 6)$, $(6, 1)$, $(2, 5)$, $(5, 2)$, $(3, 4)$, $(4, 3)$	6/36
8	$(2, 6)$, $(6, 2)$, $(3, 5)$, $(5, 3)$, $(4, 4)$	5/36
		16/36

The total, 16/36, is the probability that we win, and hence the probability that W will be $2. Since the points 6, 7, and 8 are mutually exclusive, Axiom 3 of the definition of probability (Section 1.4), permits us to calculate this probability. Clearly, $1 - (16/36) = 20/36$ is the probability that we pay $1. Hence, the distribution of W is

Values of W	Probability
$2	16/36
- $1	20/36
	1

To decide whether to play, we should calculate the expected value of this distribution:

$$E(W) = (2)\left(\frac{16}{36}\right) + (-1)\left(\frac{20}{36}\right) = \frac{12}{36} = \$0.33$$

In the long run, we expect to gain (you expect to lose) 33¢ each time we play— that's $10 in 30 games! You probably do not want to play.

If we keep your winnings at \$1, what should you pay us each time to insure a fair game? A fair game is one in which neither side has an advantage, so clearly $E(W) = 0$. Let d be the amount you pay when we win. Then in a fair game,

$$0 = E(W) = (d)\left(\frac{16}{36}\right) + (-1)\left(\frac{20}{36}\right)$$

$$\frac{20}{36} = d\left(\frac{16}{36}\right)$$

$$d = \$1.25$$

This example emphasizes the need for care in the interpretation of expected value. On any game you will either win \$1 or lose \$1.25 (under the revised rules). The zero value for $E(W)$ implies that if we play the game many times the average of our winnings, or your winnings, will be close to zero. In this case, as in many others, $E(W)$ is not one of the possible values for W.

The notion of expectation extends our discussion of odds and fair bets of Section 1.5. If we are betting on a repeatable event A, such as a full house in poker or a red stopping position in roulette, we interpret $P(A)$ as the fraction of times A occurs in a long run of trials. If we win u_2 dollars for each A outcome and lose u_1 for each outcome other than A, then in N trials we would expect $u_2 N P(A)$ dollars of winnings and $u_1 N[1 - P(A)]$ dollars of losses. Our profit and loss per trial are, respectively, $u_2 P(A)$ and $u_1 [1 - P(A)]$. In this case, we are interested in our change of assets, which we will now call our *winnings*, W, where we denote a loss as a negative win. Then, the possible values of W are

$$W = \begin{cases} u_2 & \text{with probability } P(A) \\ -u_1 & \text{with probability } 1 - P(A) \end{cases}$$

Since the outcome of any trial must be either A or not A ($\overline{A}$), our expectation for a single trial is

$$E(W) = (u_2) P(A) + (-u_1) [1 - P(A)] = u_2 P(A) - u_1 P(\overline{A}) \tag{3.1.1}$$

which corresponds to the definition of expectation given above.

We used expectation earlier to assess the fairness of a risky situation; clearly, a fair bet is one in which you can expect the same winnings no matter which side of the bet you take. Intuitively, a fair bet can occur only when expected profit and expected loss add to zero. Formally, let us compute the expectation of the change in assets for a fair bet. If X = change in assets for a fair bet, we recall from Equation (1.5.7) that

$$X = \begin{cases} Y\left(\dfrac{P(\overline{A})}{P(A)}\right) & \text{with probability } P(A) \\ -Y & \text{with probability } P(\overline{A}) \end{cases}$$

Then, according to the definition of expectation,

$$E(X) \;=\; Y\!\left(\frac{P(A)}{P(\overline{A})}\right) P(A) + (-Y)\, P(\overline{A})$$

$$ \;=\; Y\, P(\overline{A}) - Y\, P(\overline{A})$$

$$ \;=\; 0$$

Thus we can formulate the following definition.

> A *fair bet* is an EGS in which the expected value of the change in assets is zero.

Let us consider an example in this context.

Example 3.1.3 We have $10 to wager on an event A, whose probability of occurrence we have strong reason to believe is $P(A) = 1/3$. We are offered a bet where we deposit our $10 and lose it if A does not occur and where we are paid a total of $25 if A does occur. Is this a fair bet for us?

1. By Equation (1.5.8) our total assets in a fair bet would be

$$\text{Total assets} \;=\; \begin{cases} \dfrac{10}{1/3} = 30 & \text{if A occurs} \\[2mm] 0 & \text{if } \overline{A}\text{ occurs} \end{cases}$$

The zero if $\overline{A}$ occurs is matched by the offered bet, but that bet proposes $25 rather than the required $30 if A occurs. The bet is less than fair for us.

2. Since we deposit $10 and get paid a total of $25 if A occurs, losing our $10 if $\overline{A}$ occurs, our change in assets will be $25 - $10 = $15 if A occurs, and -$10 if $\overline{A}$ occurs. By equation (1.5.7) our change in assets with a fair bet should be

$$\text{Change in assets} \;=\; \begin{cases} 10\left(\dfrac{2/3}{1/3}\right) = 10(2) = 20 & \text{if A occurs} \\[2mm] -10 & \text{if } \overline{A}\text{ occurs} \end{cases}$$

The change in case $\overline{A}$ occurs is the same as in the offered bet, but that bet gives $15 rather than the required $20 if A occurs. The offered bet is less than fair for us.

3. Since $P(A) = 1/3$ and $P(\overline{A}) = 2/3$, the change (X) in assets under terms of the offered bet would be $15 with probability 1/3 and -$10 with probability 2/3. Hence, by definition of expectation given on page 61, the expected change in assets is:

$$E(X) = (15)\left(\frac{1}{3}\right) + (-10)\left(\frac{2}{3}\right) = \frac{15}{3} - \frac{20}{3} = -\frac{5}{3}$$

Since this is *negative* $1.67, the offered bet is not fair for us.

As we shall see in more detail in Chapter 5, optimizing an expected value often becomes the criterion for decision making; the following example illustrates that this may not always be the most reasonable criterion.

Example 3.1.4 Samantha, the owner of Sam's Cycle Shop, is considering various methods of advertising. She can use radio (R), television (T), newspapers (N), or mail (M); for simplicity, she wishes to concentrate all her advertising within one method. The effectiveness of each method is dependent upon the state of the economy after the ad appears. She determines her net cost (for comparable exposure) for each advertising method in each of three economic states: recession, stagflation, and inflation; this is given in Table 3.2. From economists Samantha obtains the probability distribution for states of the economy in the relevant time period and calculates the expected cost for each mode of advertising. For example, the expected cost for radio is (30)(0.1) + (60)(0.4) + (90)(0.5) = 72. The minimal expectation of 35 would lead her to use a newspaper. However, suppose Samantha cannot afford to spend more than $50 on advertising. Then she might look for the method giving the smallest probability of exceeding this limit, mail in this case.

Table 3.2 Cost Structure for Sam's Cycle Ads

Advertising methods	Costs under states of economy			Expected cost	Probability of cost > 50
	Recession	Stagflation	Inflation		
Radio (R)	30	60	90	72	0.9
Television (T)	10	40	70	52	0.5
Newspaper (N)	20	20	50	35[a]	0.5
Mail (M)	45	45	45	45	0[b]
Probability of economic state	0.1	0.4	0.5		

[a] Minimal expectation.
[b] Smallest probability of exceeding limit.

Another example (due to Hacking, 1965) illustrates the special features and limitations of expected value.

Example 3.1.5   A soft-drink machine with a capacity of 60 cans takes only quarters. The price of a can goes up from 25¢ to 30¢. Rather than replacing the machine, the supplier includes ten empty cans in such a way that one empty can is included in each batch of six consecutive cans. (This arrangement does not exclude the possibility that, say, the sixth and seventh cans may be empty.) Is this arrangement fair to the customer?

If the machine is used solely by one person, the customer will get 50 full cans during the period corresponding to a single cycle of the machine for the price of $(60)(0.25) = \$15.00$, and each full can will cost him or her on the average 30¢ (the expected number of cans for 25¢ is $(1/6)(0) + (5/6)(1) = 5/6$ cans, or equivalently one can for 30¢). The arrangement is less fair if several people use the machine. It is quite possible for one customer who purchased 10 cans (during the period corresponding to the cycle of the machine) to end up with two empty cans, while another may end up with only one; some may even get away with purchasing nine full cans with no empties, while an unlucky person may even have three empties among eight. Although these discrepancies may be ironed out during several cycles, we do not know how many cycles are needed. A casual customer who uses this machine on a single occasion might very likely find this machine unfair. There is a 5:1 chance that the can will cost him or her a quarter; there is a 1:5 chance that he or she will be required to pay 50¢ for a single can, and there is even a very small chance (3 in 100) that the can will cost him or her 75¢. In any case the customer will *never* pay the fair price of 30¢ for the can of soda.

Exercises

3.1   In playing roulette in Nevada, the house odds of 1:1 for a bet on red mean that for every dollar you bet on red you will be paid a dollar (in addition to return of your bet) if red occurs, and will lose your dollar if red does not occur.
    a.   What are your expected winnings for each game?
    b.   Is the bet fair?

3.2   A prisoner is given 10 white balls, 10 black balls, and two boxes. He is told that an executioner will draw one ball from one of the two boxes. If it is white, the prisoner will go free; if it is black, he will die. How should the prisoner arrange the balls in the boxes to give himself the best chance for survival?

3.3   Will you be pleased to bet $25 on event A provided that if the event occurs you get paid a total of $100 (and if the event does not occur you lose your $25), given that your assessment of the likelihood of A is $P(A) = 1/5$? Consider:

    a.   total assets

    b.   change in assets

    c.   expected change in assets

3.4   Are you willing to bet $2 that point 7 occurs on the next roll of a pair of dice, provided that you win $9 if the 7 actually occurs (and lose your $2 if it does not)? Compute the actual expected change in assets. Determine the odds which are being proposed by the house, and compare with fair odds.

3.5   A fair coin is tossed until a head comes up. What is the expected number of tosses required? [Hint: Let X = number of tosses required. First, find P(X = 1) = P(head on first toss); similarly, compute P(X = 2), . . . , P(X = k).]

3.6   An item on sale costs $75.00 and it may be defective with probability 0.3. A guaranteed item costs $100.00. Which should you buy?

3.7   The mean IQ of the population of eighth-graders in a city is known to be 100. You have selected a sample of 50 children for a study of educational achievement. The first child tested has an IQ of 150. What do you expect the mean IQ to be for the whole sample?

3.8   Let $\bar{X}_W$ represent an average of certain characteristics for whites and $\bar{X}_B$ the average of the same characteristic for blacks. Let N_W and N_B be the number of whites and blacks, respectively. Find the average for the total population: $\bar{\bar{X}}$.

3.9   In a certain Zambian village live 600 women. Seven percent of them wear one earring. Of the other 93 percent, half are wearing two earrings, half are wearing none. What is the expected number of earrings worn by each woman? What is the expected number of earrings being worn in the village?

3.10   Using the complete probability distribution of the point rolled with a pair of fair dice given in Appendix Table B.1, compute the expected value of the point rolled. (Hint: Making the calculation is most convenient if probabilities are taken in fractional form with 36 as denominator.)

3.11   Your club has a fund-raising event scheduled for a week from today; the activity, depending on the weather, will be either an outdoor picnic or an indoor flea market. Today you must decide and send out the final notice. Your experience has been that in case of no rain the picnic will yield a profit of $400 while the flea market will yield only $150, but that in case of rain the picnic produces a profit of $100 while the flea market produces $300. The most dependable prediction you can get is a 40 percent probability of precipitation for the scheduled day. If we use expected profit as a criterion, what decision should we make? For what probability (say p) of precipitation would the expectation be the same for both events?

3.12   A small village bakery anticipates the following distribution of demand X this coming Fourth of July for cheesecakes, a holiday specialty: $P(X = 0) = 0.1$, $P(X = 1) = 0.2$, $P(X = 2) = 0.3$, $P(X = 3) = 0.2$, $P(X = 4) = 0.1$, $P(X = 5) = 0.1$.
   a.   What is the expected value of the demand?
   b.   Suppose that the bakery makes a net profit (G for gain) of $5 on each cake sold but loses $6 ($G = -6$) for each cake left unsold. If the bakery wishes to maximize expected profit, how many cakes should be prepared? What is the maximum number that should be made if the bakery uses the principle of doing all it can just as long as its expectation of profit is not negative? [Hint: If the bakery prepared one cake, $E(G) = (-6) P(X = 0) + (5) P(X \geqslant 1)$, where $P(X \geqslant 1) = P(X = 1) + P(X = 2) + P(X = 3) + P(X = 4) + P(X = 5)$, since the one cake will be sold if the demand is one or more. If the bakery prepares two cakes, $E(G) = (-12) P(X = 0) + (5 - 6) P(X = 1) + (10) P(X \geqslant 2)$. Continue for 3, 4, 5 cakes prepared, and compare the various expected values.]

3.13   Let p be the probability that a woman living in sexual union conceives in any given month.*
   a.   What is the probability of conception occurring in the fourth month and not before? (Hint: If C_i is the event "conception in the ith month," then the event of interest is $\bar{C}_1 \cap \bar{C}_2 \cap \bar{C}_3 \cap \bar{C}_4$.)
   b.   Generalize the result of question a for the kth month.
   c.   What is the average (expected) waiting time until conception?

3.14   You are standing outside the gates of the Bluebonnet Bowl with four extra tickets to the game. You bought the tickets for $7.50 each. You quickly determine the following probabilities for selling the excess tickets for $25 each. You cannot get a refund for any unsold tickets.

x = no. of excess tickets sold	$P(X = x)$
0	0.10
1	0.20
2	0.30
3	0.30
4	0.10

Find the expected value of your profit.

3.15   You have just purchased an auto insurance policy for your new car from Bond Insurance Co. for $500. The probabilities of a major collision and minor collision during the first year of operation are 0.05 and 0.15, respectively. The policy will pay you $2000 for a major collision or

---

*Statistics show that for women in the age range 15–45 the average value of $p = 0.1$; for women in the prime childbearing ages of 20–35 the value of p is about 0.25.

$600 for minor collision, should one occur. The policy also pays a dividend of $100 for those who have no collisions during the year. Assuming you have at most one collision in the coming year, what is Bond Insurance Co.'s expected profit on your policy?

3.16   *Fiesta Holiday* (*Los Angeles Times*, September 2, 1970). During the late 1960s and into 1970 the Hacienda Hotel had the following advertisement in the *Los Angeles Times*:

FIESTA HOLIDAY
2 DAYS/1 NIGHT FOR ONE OR TWO PERSONS
You Pay $15
WE GIVE YOU $10 in FREE play chips!
Your NET Cost is $5

Upon arrival one learns that the $10 in "free" play chips are 10 chips to be used in the following manner during gambling. If one wins when playing such a chip, one receives exactly $1 and must relinquish the chip; of course, if one loses, the chip must still be relinquished. This misrepresentation resulted in a judgment of $1,250,000 in a lawsuit. Explain the probabilistic grounds for this decision by comparing the advertised net cost of $5 with expected net cost. [Hint: Let p be the probability of winning with any one chip, and assume $p < 0.5$, which is the case with all commercial games of chance (see Section 3.2).]

3.2 GAMES OF CHANCE

> *We figured the odds as best we could, and then we rolled the dice.*
> Jimmy Carter

It is not uncommon for people to try to control chance, usually through superstition. We have all found ourselves playing a hunch, interpreting a coincidence or a dream, or falling prey to the gambler's fallacy (Section 1.3). Modern probability theory originated in the middle of the seventeenth century in France from discussions of the mathematicians Pierre de Fermat (1601–1665) and Blaise Pascal (1623–1662) about games of chance involving dice. Big-time gambling interests benefit from these hunches, while the odds of the games, just as insurance rates, are based on the rules of probability, which tame the unpredictable behavior of chance. Thus, insurance is usually helpful and gambling is sometimes fun. Let us see why.

Example 3.2.1 Las Vegas Wheel of Fortune   A typical Las Vegas wheel of fortune (or big six wheel) has 54 positions: one joker, J, one flag, F, two 20s, four 10s, seven 5s, fifteen 2s, and twenty-four 1s. (The model used in many Atlantic City and Nevada casinos is a six-foot vertically mounted wheel with horizontal, equally spaced pegs in its rim. As the wheel spins, a rubber flapper strikes successive pegs, slowing the wheel.) If we bet \$1 on J and J comes up, we win \$40. Similarly, if we bet \$1 on the flag and the flag comes up, we win \$40.

Let W denote our winnings for this game. If we are equally likely to bet on any of the 54 positions, the discussion of Section 3.1 allows us to compute our expected winnings:

$$E(W) = 40 \left(\frac{1}{54}\right) - 1 \left(\frac{53}{54}\right) = -0.24$$

We see that the game is not fair, or that we are overpaying for our chance to play. What should our bet be for a fair game? Clearly, an amount x making

$$E(W) = 40 \left(\frac{1}{54}\right) - x \left(\frac{53}{54}\right) = 0$$

That is,

$$x = \left(\frac{40}{54}\right) \left(\frac{54}{53}\right) \approx 0.75$$

It should not be surprising that for all commercial games of chance (with the exception of a version of blackjack known as "21," which is not a game of pure chance) the expected win is always less than zero, while the expected win in a gambling procedure under indifference is always equal to zero. The difference between these two values can be regarded as compensation for entertainment. Hence, according to the results developed in Sections 1.5 and 3.1, we should be displeased betting on any one of the commercial games of chance (with the exception of the blackjack game). Of course, our rational theory of Section 1.5 ignored the entertainment (or the excitement as well as the danger) involved in gambling.

Turning the tables, the proprietor of the gambling establishment is interested in the average profit of the house on each dollar bet: the ratio of the expected win to the amount bet. This is a negative ratio in most cases, so the value is multiplied by -1 to make it positive and by 100 to report it as a percentage. Thus we have the definition:

$$\text{House percentage or house advantage} = \frac{\text{expected win}}{\text{amount of bet}} (-100)$$

For position J on the wheel of fortune, the house percentage is

$$\frac{40(1/54) + (-1)(53/54)}{1} (-100) = 24.1\%$$

On the average, the house collects 24¢ pure profit on each $1 bet on J—quite a healthy profit! Details on other bets are given in Exercises 3.17 to 3.19.

Example 3.2.2   Nevada-Style Roulette   This consists of a wheel which has 38 positions (grooves), with 36 positions marked consecutively by the numbers 1 to 36 and two other positions marked "0" and "00," respectively. The wheel is in continuous motion and a small metal ball is set in motion at the rim of the wheel, rotating in the opposite direction. Eventually the ball drops into one of the grooves; the number marked on this groove is the winning number. Bets are made by placing chips on a layout near the wheel (Figure 3.1). Players may make bets in several different ways. We describe a few of these:

1. In a *straight bet* we bet on any single number, including 0 and 00, by putting a dollar chip on that particular number. If this number comes up, we win $35 and take our chip back; otherwise we lose the chip.
2. In a *split bet* we put a dollar chip on the line between two numbers. If either number comes up, the payoff is $17 and the chip is returned.
3. By placing a chip on the *first dozen* location on the betting layout we receive a payoff of $2 plus the initial chip if any of the numbers 1-12 comes up.
4. We may bet on *odd* and receive $1 plus the initial chip if any odd number between 1 and 36 comes up. Similarly, we may bet on even, red, or black (separately) with the same rewards.

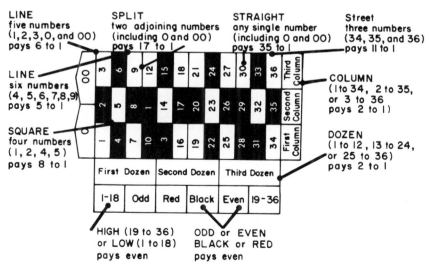

Figure 3.1   Single chip bets in Nevada roulette.

Let us consider the characteristic of the odd (equivalently, even) game. Let A be the event of a win in this game: then $P(A) = 18/38$ and $P(\overline{A}) = 20/38$. The expected win is therefore

$$(1) \left(\frac{18}{38}\right) + (-1) \left(\frac{20}{38}\right) = \frac{-2}{38} = -5.26\%$$

From Equation (1.5.7), the indifference win (change in assets) is

$$Y \cdot \frac{P(\overline{A})}{P(A)} = \frac{(1)(20/38)}{18/38} = \frac{20}{18} = \$1.11$$

which is greater than the actual win of $1. The indifference bet corresponding to a win of $1 is given by x where, from Equation (1.5.8),

$$\frac{x}{18/38} = 2 = \text{total assets in case of a win}$$

hence, $\quad x = \dfrac{18}{19} = \$0.95$

which is less than our original $1 bet. The house percentage of the odd (even) game is

$$\frac{-2/38}{1} \ (-100) = 5.26\%$$

a relatively moderate profit.

In Exercise 3.22 the reader is asked to compute the expected win, the indifference win, the indifference bet, and the house percentage for other single chip bets on Nevada roulette. You will find that house percentages are equal for all bets except in the case of a line bet on five numbers, when the house percentage is 7.89 percent. Stay away from five numbers!

Example 3.2.3 Game of Craps (Dice) Bernard de Mandeville brought the dice game "craps" to New Orleans in 1813; its name is derived from the nickname for a Creole, "Johnny Crapaud." One of the reasons for the popularity of this game of chance is that the house percentage is below 1.5 percent. A symmetric half of the table layout is given in Figure 3.2. We bet on the *pass line* by placing our (dollar) chips on it, and the shooter (usually one of the players with the largest bet) throws two dice. The rules for payoff are as follows: If the sum of numbers on two dice (points) is 7 or 11, the shooter and players who bet with him or her on pass line win; the player loses on the first throw if the shooter gets 2, 3, or 12 points. If the outcome is 4, 5, 6, 8, 9, or 10 points on the first throw, the shooter continues to throw the dice repeatedly until he or she produces either a 7 or the first number thrown; in the latter the player wins, in the former the player loses. If the player loses, the croupier (attendant) takes the chip. If the

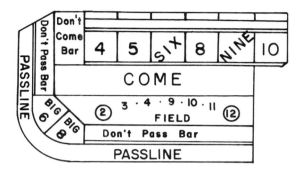

Figure 3.2   Casino craps layout (left side).

player wins, the croupier places a chip next to the player's chip and the player may now pick up both chips. (Note that number 7 plays here the double role of a winner in the beginning and a loser at the end.) Examples of throws are as follows:

5	8	2	3	4	6	5	(win)
8	4	2	12	10	11	7	(loss)
12							(loss)
7							(win)

To evaluate the probability of winning the pass line version of the game of craps with a pair of fair dice, we first compute from Chapter 1 the probability of various points with a pair of dice, summarized in Table 3.3. From this table the probability of winning on the first toss is

$$P(7) + P(11) = (6/36) + (2/36) = 8/36 = 2/9$$

If the first toss is 4, we shall win if the next toss is also 4. We shall also win if the first three outcomes are 4N4, where the letter N represents a number *different* from 4 or 7. We shall win if the first four outcomes are 4NN4. In general, we shall win if the first k outcomes are $\underbrace{4N \ldots N4}_{k-2 \text{ times}}$, with k = 2, 3, ...

Table 3.3   Probability of Various Points in Pass Line Craps

			Point								
	2	3	4	5	6	7	8	9	10	11	12
Probability	1/36	2/36	3/36	4/36	5/36	6/36	5/36	4/36	3/36	2/36	1/36

The probability of such a sequence is

$$\left(\frac{3}{36}\right)\underbrace{\left(\frac{27}{36}\right)\cdots\left(\frac{27}{36}\right)}_{k-2 \text{ times}}\left(\frac{3}{36}\right)$$

and the probability of winning with the first toss of 4 is thus

$$\left(\frac{3}{36}\right)\left(\frac{3}{36}\right) + \left(\frac{3}{36}\right)\left(\frac{3}{36}\right)\left(\frac{27}{36}\right) + \left(\frac{3}{36}\right)\left(\frac{3}{36}\right)\left(\frac{27}{36}\right)^2$$

$$+ \left(\frac{3}{36}\right)\left(\frac{3}{36}\right)\left(\frac{27}{36}\right)^3 + \cdots + \left(\frac{3}{36}\right)^2\left(\frac{27}{36}\right)^k + \cdots$$

This is a sum of geometric progression with first time $(3/36)^2$ and ratio $(27/36)$ and has the value

$$S = \frac{\text{first term}}{1 - \text{ratio}} = \frac{(3/36)^2}{1 - (27/26)} = \frac{1}{36}$$

Identical calculations with $(3/36)^2$ and $27/36$ replaced by $26/36$ show that the probability of winning on 5 is

$$\frac{(4/36)^2}{1 - (26/36)} = \frac{2}{45}$$

and on 6 is

$$\frac{(5/36)^2}{1 - (25/36)} = \frac{25}{396}$$

The probability of winning on 8 is the same as the probability of winning on 6, the probability of winning on 9 is the same as on 5, and on 10 is the same as on 4. Thus the probability of winning on the pass version of the game of craps is

$$\sum_{k=4}^{11} P(\text{win when first toss is k}) = \frac{2}{9} + 2\left(\frac{1}{36}\right) + 2\left(\frac{2}{45}\right) + 2\left(\frac{25}{396}\right)$$

$$= \frac{976}{1980} = \frac{244}{495} = 0.4929$$

slightly less than 1/2. The expected win on $1 bet is thus:

$$(1)\left(\frac{244}{495}\right) + (-1)\left(\frac{251}{495}\right) = -\frac{7}{495}$$

and the house percentage is

$$(-100)\left(\frac{-7/495}{1}\right) = 1.41\%$$

An alternative method of calculating these probabilities is by using the method of reduced sample space. If the first outcome of a sequence of shots is 4, then the outcome of this sequence is determined only by whether 7 appears before another 4 (in which case we lose) or whether another 4 appears first (before 7), in which case we win. We thus have a *conditional* space of nine possibilities (six for 7 and three for 4). Hence the probability of getting 4 before 7 is 3/9 = 1/3, and the probability of winning on the outcome 4 is equal to probability of having a 4 at the first place (3/36) times the probability of 4 before 7 (1/3); this is equal to 1/36, the value obtained using the geometric progression method. Similarly, the probability of winning on 5 is (4/36)(4/10) = 2/45 and the probability on 6 is (5/30)(5/11) = 25/396, which coincide with the results above.

It is interesting to investigate probabilistically the average duration (i.e., the average number of throws) on this game. The probability that the game ends in exactly N throws (for N = 1, 2, 3, 4, or 5) is given in Table 3.4. The probability that the game will end on the first or second throw is (1/3) + (244/1296) = 0.523, greater than 1/2.

The calculations of Table 3.4 were carried out under the assumption that both dice are fair. The notion of a fair die is a hypothetical one; one cannot give a proof in the mathematical sense that a particular die is a fair one. Therefore, it is worthwhile to investigate the variation in the probability of winning craps when the dice are biased. Probabilities can be changed in many ways, and we shall consider only one case. Suppose that both dice are biased

Table 3.4 Probability Distribution for the Duration of Craps

N, duration	Probability game ends in N throws
1	$P(\text{outcome is } 7, 11, 2, 3, \text{ or } 12) = \dfrac{6}{36} + \dfrac{2}{36} + \dfrac{1}{36} + \dfrac{2}{36} + \dfrac{1}{36}$
	$= \dfrac{12}{36} = \dfrac{1}{3}$
2	$2\left(\dfrac{9}{36}\right)\left(\dfrac{3}{36}\right) + 2\left(\dfrac{10}{36}\right)\left(\dfrac{4}{36}\right) + 2\left(\dfrac{11}{36}\right)\left(\dfrac{5}{36}\right) = \dfrac{244}{1296} = 0.190$
3 . . .	$2\left(\dfrac{27}{36}\right)\left(\dfrac{9}{36}\right)\left(\dfrac{3}{36}\right) + 2\left(\dfrac{26}{36}\right)\left(\dfrac{10}{36}\right)\left(\dfrac{4}{36}\right) + 2\left(\dfrac{25}{36}\right)\left(\dfrac{11}{36}\right)\left(\dfrac{5}{36}\right) \approx 0.134$
n	$2\left(\dfrac{27}{36}\right)^{n-2}\left(\dfrac{9}{36}\right)\left(\dfrac{3}{36}\right) + 2\left(\dfrac{26}{36}\right)^{n-2}\left(\dfrac{10}{36}\right)\left(\dfrac{4}{36}\right) + 2\left(\dfrac{25}{36}\right)^{n-2}\left(\dfrac{11}{36}\right)\left(\dfrac{5}{36}\right)$

Table 3.5   Distribution of Outcomes for a Single Biased Die

	Face					
	1	2	3	4	5	6
Probability	2/15	2/15	2/15	1/5	1/5	1/5

toward the large numbers, i.e., the probabilities for each face of a die are not 1/6 but are those given in Table 3.5. In this case throws of small numbers (1, 2) yield losing points of 2, 3, and 4, and large numbers (5, 6) are more likely to yield outcome 11, a winning point. Intuitively, then, the probability of winning has increased (to greater than 1/2) with dice biased in this way. The point 7 plays an important role in craps: for fair dice it is the most likely value, the only value which can occur regardless of which outcome occurs on one die, and it guarantees a win if thrown on the first toss and a loss if thrown later. In fact (see Exercise 3.35), increasing the probability of 7 while keeping other probabilities proportional decreases the probability of winning. Now the biased dice defined by Table 3.5 increase the probability of winning, since the dice are biased toward the large numbers and are of the wear pattern which favors parallel rather than opposite numbers (totaling 7). In Exercise 3.34 the reader is required to verify that for these two dice the probability of winning is 0.522.

Betting Systems

Various systems for roulette and craps supposedly guarantee a win, at least over a substantial number of rounds. All of these are based on a combination of bets (a *compound bet*) with different stakes for the individual bets. These systems are evidently without any foundation, since any combination of bets with negative expectations (i.e., positive house percentages) results in a bet with a negative expectation. (Any average of negative numbers is again a negative number.) Exercise 3.44 asks you to analyze these systems carefully and compute actual house percentages in each of these compound bets.

Roulette.   Since the third column of a roulette betting layout includes eight red numbers and only four black numbers (see Figure 3.1), the player is told to bet $1.00 (or some other fixed amount) on the third column and the same amount on the red. The motivation for such a bet is as follows: In 38 rounds, 0 and 00 appear on the average once, resulting in a total loss of $4.00. Red appears on the average 18 times, of which 8 will occur in the third column (and 10 in other columns), yielding a loss of $2.00 10 times and a win of $2.00 8 times (recall that a column bet pays 2:1). The resulting net loss is $20.00 – $16.00 = $4.00. Black appears 18 times. Fourteen of these will occur in the first and second columns, in which case one bet is a loss but the other is a win, so the result is an even break; four times black will appear in the third column, yielding a net gain of $(1)(\$4.00) + (2)(\$4.00) = \$12.00$. Consequently, the total change in assets in the course of 38 rounds will be $-\$4.00 - \$4.00 + \$12.00 = \$4.00$.

Compound Bet in Pass Line Craps. Place $10 on point 6, $10 on point 8, $5 on a field bet (2, 3, 4, 9, 10, 11, 12) and $1 on point 5. (See Exercises 3.28 and 3.29 for an explanation of these bets.) The motivation is that you win in all cases except when point 7 comes out, and the probability of point 7 is only 1/6. However, this motivation is misguided: we compute the house percentage for this bet for one roll, and Exercise 3.44b asks you to compute the percentage for two rolls. If 2 or 12 occurs (probability 2/36), we win $10 and the point bets remain untouched. If 3, 4, 9, 10, or 11 occurs (probability 14/36), we win $5. Since the bet on 5 pays 7:5, if 5 occurs, we win $7 but lose our field bet, resulting in a net gain of $2. Since the bet on 6 or 8 pays 7:6, with probability 10/36 we win $1 but lose the $5 field bet, with a net gain of $6. If 7 occurs, we lose $30. Thus our expected winnings are

$$
\begin{aligned}
E(W) &= (2/36)(10) + (14/36)(5) + (4/36)(2) \\
&\quad + (10/36)(5) + (6/36)(-30) \\
&= -22/36 = -11/18
\end{aligned}
$$

The house percentage is therefore $(1100/18)/30 = 2.03$ percent.

Don't Pass Craps. Place $35.00 on the don't pass line; if a second roll is required, place $10.00 on 6, $10.00 on 8, $5.00 on the field bet and $5.00 on point 5. Take your bets down after the second roll. The motivation for this bet is that we win $5.00 on each roll except when the shooter repeats his or her number in two rolls. In Exercise 3.44c you are asked to show that the house percentage for this betting system is a whopping 7.42 percent. This is because the advantages of the don't pass line bets and the point bets are not utilized by taking down your bets after the second roll.

We conclude this section with a remark about the new electronic slot-machine games, which are becoming more and more prevalent. Unlike the standard mechanical slot machines at all casinos, electronic games operate on the principle of random selection of numbers without replacement. One of the most popular games, Draw-a-String, selects sequentially without replacement from 36 available numbers. There are several winning outcomes (based on strings of consecutive numbers) and only one losing outcome; there is also a small chance to win a large amount. Surprisingly, the house advantage is about 7 percent, well below the standard mechanical slot machine. Detailed descriptions of the game as well as computations for the payoff and house percentage are given in a recent paper by Johnson and Smith (1981).

In the long run, you just cannot beat the house in a game of chance! The exercises that follow this section supply as much information as possible about some common games of chance. Even if you do not feel inclined to solve each one, at least read them before you visit a casino.

Exercises

Wheels

3.17   Compute your expected winnings and ticket values under indifference for the Las Vegas wheel of fortune in the following game: a bet of $1 on the 20 pays $20, on the 10 pays $10, on the 5 pays $5, on the 2 pays $2, and on the 1 pays $1.

3.18   Show that the house advantage for the wheel of fortune for position
   a.   20 is 22.2%
   b.   10 is 18.5%
   c.   5 is 22.2%
   d.   2 is 16.67%
   e.   1 is 11%

3.19   On the basis of Exercises 3.17 and 3.18, justify the statement: If you are compelled to bet on a wheel of fortune, bet on 1, and stay away from jokers, 20s, and 5s.

3.20   A multicolor wheel consists of 25 compartments: 6 green, 6 red, 6 white, 6 yellow (each color numbered 4, 3, 3, 3, 2, 2), and 1 blue (numbered 24). If a ball comes to rest on a selected color (green, red, white, or yellow) the player is paid 4:1, 3:1, or 2:1 determined by the number marked on the specific compartment at which the ball stops. If a player bets on blue and blue occurs the player wins 24:1. Calculate the house percentage for bets on various colors.

3.21   *La Boule* involves a wheel subdivided into 18 segments marked in a clockwise direction 1, 2, ... , 9 and 1, 2, ... , 9. The segments numbered 1, 3, 6, and 8 are black and those numbered 2, 4, 7, and 9 are red. A single-number bet pays 7:1, multiple bets of the type black, red, even, and odd (the last excluding number 5) pay 1:1. Compute the house percentage of the single bet and of multiple bets for this game.

Roulette

3.22   For Nevada roulette compute the expected win, the indifference win, the indifference bet, and the house percentage for the following single-chip bets shown in Figure 3.1, with the corresponding odds.
   a.   Street bet: a $1 chip is placed on the line at the side of three numbers.
   b.   Square line bet: a $1 chip is placed at the center of four numbers.
   c.   Line bet: a $1 chip is placed on the side of the dividing line between five or six specific numbers.

d. Dozen bet: a \$1 chip is placed on the square designated first dozen (1–12), second dozen (13–24), or third dozen (25–36).

e. Column bet: a \$1 chip is placed at the end of one of three columns, 1–34, 2–35, or 3–36.

f. Red or black, high or low, odd or even, placed as indicated in Figure 3.1.

3.23 Consider the following compound bet in roulette. A player bets \$M that the ball will land on 17, 18, 20, or 21 and also \$D that the ball will land on an even number. Compute the house percentage. [Hint: Distinguish four cases: (1) 17 or 21; (b) 18 or 20; (c) even number other than 18 or 20; and (d) 0, 00, or odd number other than 17 or 21. Compute probabilities and value of winnings in each case.]

3.24 A Monte Carlo roulette wheel has only one 0 (and thus 37 positions). If you bet on a red number, say, and 0 turns up, you don't lose your bet but you "go to prison." This means that you continue to participate in the game. If the next round yields red, your bet is returned; if the next round yields black, you lose the bet; and if the next round is 0 again, you continue your participation under the rules just described. Show that the house percentage of this game is only 1.35 percent. (Use either the method of reduced samples or the geometric progression as in the computations of the house percentage for pass line craps.)

3.25 In Atlantic City casinos the outside bets in roulette (high or low, odd or even, black or red) are played according to the following rules: If the right number or color comes up, the player wins the amount of his or her bet; if the wrong number (or color) occurs, the player loses his or her bet; if 00 or 0 occur, the player loses one-half of his or her bet. Calculate the house percentage.

Craps and Dice

3.26 Verify that for the pass line version of the game of craps, the probability of the game ending by the
a. first three throws is 0.656
b. first four throws is 0.753
c. first five throws is 0.822

3.27 *Don't Pass Line Version of Craps.* If the shooter gets 2 or 3 points the player wins; if the shooter gets 7 or 11 the player loses; if the shooter gets 12 nobody wins. The points 4, 5, 6, 8, 9, and 10 require continuation of the game until the first occurrence of 7, in which case the player wins, or the first recurrence of the point originally obtained, in which case the player loses. This game is in a certain sense a mirror image of the pass line version; however, there is a positive probability of tie: if 12 occurs, you may pick up your bet and walk away. Compute the probability of winning in this version of craps, the expected win on a \$1 bet, and house percentage. [Hint: The sample space may be viewed in either of two ways; ignore the outcome 12 and have 35 outcomes, or include it and have the usual 36. To compute the probability of winning, observe that there are 7 ways to win: (1) 2 or 3 on first throw; (2) getting 5 and then 7 before 5; (3) getting 4 and then 7 before 4; (4) getting 6 and then 7 before 6; (5) getting 8 and then 7 before 8; (6) getting 9 and then 7 before 9; and (7) getting 10 and then 7 before 10.]

3.28   In a *big 6* or *big 8* bet placed at craps you win one dollar if point 6 occurs before point 7, and you lose one dollar otherwise. Compute the house percentage for this bet.

3.29   You place a dollar chip on the craps layout where "field" is marked. If 2 or 12 comes up, you win two dollars (and the bet is returned). If 3, 4, 9, 10 or 11 comes up, you win one dollar (and your bet is returned). If 5, 6, 7, or 8 comes up, you lose your bet. Compute the house percentage in this case.

3.30   *Buying the numbers* is a version of craps in which you "buy" the number (or point) 5 by placing \$20 on this number and paying \$1 to the house. If 5 comes before 7 you win according to the correct odds (1:2), i.e., you are paid \$30; otherwise, you lose your bet and your payment. Calculate the house percentage.

3.31   In *any seven*, a version of craps, two dice are tossed only once. If point 7 occurs, your payoff is 4:1; otherwise, you lose your bet and your payment. Calculate the house percentage.

3.32   a.   In playing *hardways 4* in craps, you are betting that a 4 will come up as a 2 + 2. If it comes up as 1 + 3 or 3 + 1, or if a 7 comes up, you lose. If you do get a 2 + 2 before any other combinations of 4 or before a 7, your payoff is 7:1. Calculate the house percentage of this game.

       b.   In *hardways 10* you are betting that the dice will show 5 + 5 before any other combination of 10 or a 7. If the payoff is the same as hardways 4, what is the house percentage?

3.33   Obtain the value for the mathematical expectation of the number of throws N in a round of pass line craps. Proceed as follows: Compute the value of $1(12/36) + 2(244/1296) + 3P(N = 4) + \cdots$, where $P(N = n) = 2(27/36)^{n-2} (9/36)(3/36) + 2(26/36)^{n-2} (10/36)(4/36) + 2(25/36)^{n-2} (11/36)(5/36)$. In this sum, we see the following expressions:

$$3(27/36) + 4(27/36)^2 + 5(27/36)^3 + \cdots$$
$$3(26/36) + 4(26/36)^2 + 5(26/36)^3 + \cdots$$
$$3(25/36) + 4(25/36)^2 + 5(25/36)^3 + \cdots$$

Multiply each one of these expressions by the corresponding quotient to get expressions of the form $E = 3a^2 + 4a^3 + 5a^4 + \cdots$ where $a = 27/36,\ 26/36,\ 25/36$, respectively. Compute aE; then subtract the expression for E from the expression for aE and use the formula for geometric progression. If you are careful in your algebra, you will obtain the following expression for the expected value of N: $(12/36) + (1/6) [1 + (9/36)^{-1}] + (2/9)[1 + (10/36)^{-1}] + (5/18)[1 + (11/36)^{-1}]$.

3.34   For the biased dice described in Table 3.5, show that the probability of winning pass line craps is greater than 1/2. (Hint: construct a table of probabilities for points resulting from these dice.)

3.35   Consider the dice biased as follows:

	Face					
	1	2	3	4	5	6
Probability for die 1	2/15	2/15	2/15	1/5	1/5	1/5
Probability for die 2	1/5	1/5	1/5	2/15	2/15	2/15

Show that the probability of winning in pass line craps with these dice is less than with fair dice.

3.36   Assume the following hypothetical pair of dice. Let probability of point 7 for this pair of dice be p (a number between 0 and 1) and the probabilities of other points be proportionally

Probability of 2 = probability of 12 = $(1 - p)/30$
Probability of 3 = probability of 11 = $2(1 - p)/30$
Probability of 4 = probability of 10 = $3(1 - p)/30$
Probability of 5 = probability of 9 = $4(1 - p)/30$
Probability of 6 = probability of 8 = $5(1 - p)/30$

Applying the method of reduced sample spaces, show that the probability of win in the pass version of the game of craps is equal to

$$\left(\frac{1}{15}\right)\left[1 + 14p + (1 - p)^2 \left(\frac{5}{1 + 5p} + \frac{8}{2 + 13p} + \frac{3}{1 + 9p}\right)\right]$$

Using this result, show that the probability of win for p = 0.2 is less than the probability for p = 1/6 ($\cong$ 0.167), which corresponds to the case of a pair of fair dice. [It can be shown (see Bryson, 1973) that such dice exist only if the value of p = 1/6, i.e., only if the dice are fair.]

3.37   *Chuck-a-luck* is played with three dice of different colors, which are tumbled around in a cage. You may bet ($1, say), on one of the numbers 1, 2, 3, 4, 5, or 6; suppose you bet on 6. The cage is inverted and the dice fall on the table. If all three dice show 6 you win $3; if exactly two show 6, you win $2. If one shows 6, you win $1. Otherwise you lose your dollar. Compute the house percentage. (Hint: The probability that all three dice show 6 is 1/216. What is the probability that two dice show 6?)

3.38   In *high dice* the banker and then the player each throw a pair of dice. The player wins if he or she throws the higher number; otherwise, the player loses. Calculate the bank's advantage.

3.39   In *modified high dice* the rules of Exercise 3.38 hold, except a player's throw of 2 wins over any throw of the banker. Calculate the bank's advantage.

3.40   One die is so loaded that the probability of an ace is 1/5, other sides are equally likely. A second die is so loaded that the probability of 6 is 1/5, other sides are equally likely. What is the probability of getting point 7 with these two dice?

3.41   Two dice are rolled successively until at least one of them shows the outcome 6. What is the probability that the first throw which achieves 6 contains exactly one 6 (i.e., only one die shows 6). (Hint: Use the method of reduced samples.)

3.42   Generalize Exercise 3.41 to the case of three dice and find the probability that the first throw which achieves at least one 6 contains exactly two 6s.

3.43   Three dice are rolled in an attempt to achieve the outcome 6 on all the dice. Following each throw, those showing 6 are set aside and the remainder are thrown again; this process is continued until all three dice show a 6. What is the expected number (E_3) of throws? [Hint: The expected number of rolls to obtain at least one 6 is $6^3/(6^3 - 5^3) =$ 216/91. If the throw which achieves the *first* 6 contains no other 6 (with probability 75/91), we then have the problem with two dice (with the expected number of throws E_2). If the throw which achieves the first 6 contains exactly one more 6 (with probability 15/91) we have a problem with one die (with the expected number of throws $E_1 = 6$). Hence, E_3 is $(216/91) + (75/91)E_2 + (15/91)E_1$.]

3.44   a.   What is the actual house percentage for the fallacious roulette betting system discussed in the text? (Hint: Consider two possibilities, black bet or red bet.)
        b.   Carry out the computations for the pass line craps compound betting system for two rolls and verify that the house percentage in this case is 4.07 percent.
        c.   Show that the house percentage is 7.42 percent for the compound system on don't pass craps.

3.45   Player A rolls n + 1 dice and keeps the highest n. Player B rolls n dice. The higher total wins, with the ties awarded to Player B.
        a.   For n = 1, find the probability that Player A wins.
        b.   For n = 2, find the probability that Player A wins. [Hint: Let X_1, X_2, X_3, Y_1, Y_2 be outcomes of rolling five dice and let $U = X_{(2)} + X_{(3)}$ and $V = Y_1 + Y_2$, where $X_{(2)}$ and $X_{(3)}$ are the two largest among X_1, X_2, X_3. Compute joint probabilities for U and V by direct enumeration of cases and add probabilities to find $P(U > V)$.]

Cards

3.46   In the game of *faro* two cards are drawn from a regular deck of cards; if neither is a queen we wait for the next two cards to be drawn, and so on. If the first is not a queen but the second is, you win $1; if the first is a queen but the second is not, you lose $1; if both are queens the house collects half of your dollar bet (split). Using the method of odds, compute the house percentage for faro, assuming that the composition of the deck does not change.

3.47 In a casino version of blackjack, two cards are dealt to a player (face down) and two cards to the dealer (one face down and one face up). If the dealer's face-up card is an ace, the player is invited to take insurance against the possibility that the dealer has blackjack (i.e., that the dealer's face-down card is a ten). If it turns out that the dealer has blackjack the player is paid twice the amount of his or her bet; otherwise the player loses the insurance bet. Compute the house percentage of the insurance bet if

a. neither of the player's cards is a ten
b. exactly one of the player's cards is a ten
c. both of the player's cards are tens

(See also Exercise 2.10.)

Coins

3.48 Two people toss a coin in succession. If the outcomes are different, heads wins both coins. If the outcomes are the same, it is a tie.

a. What is the probability of a tie?
b. What is the probability of a win?
c. Suppose the first person has x coins and the second person has y coins. Let $E(x, y)$ denote the average number of tosses that will occur before one of them goes broke. Compute $E(1, 1)$, $E(2, 1)$, $E(1, 2)$, $E(2, 2)$. [Hint: Verify $E(x, y) = E(x + 1, y - 1) + E(x - 1, y + 1)$.]

3.49 Three players designated as A, B, and C play *rotation*. We first match A against B, and the winner plays C. The winner of the second match then plays the loser of the first match and the players are rotated according to this procedure until one player wins twice. If the players are of equal strength, what are the probabilities P_A, P_B, and P_C that players A, B, and C, respectively, will win? [Hint: Let p_w be the probability of players winning the game immediately after winning a single match and p_ℓ the probability of winning eventually after losing a match. Verify that $P_A = P_B = (1/2)p_w + (1/2)p_\ell$, $P_C = (1/2)p_w$, and $p_w = 1/2 + (1/2)p_\ell$.]

3.50 Consider the following game of chance. A player is to toss a coin and receive 1, 4, 9, 16, ..., n^2 dollars if the first head comes up on the first, second, third, fourth, ..., nth toss, respectively. What is the value of this game?

[Hint: Use the method of Exercise 3.33 with a = 1/2. Formally you are requested to evaluate the sum

$$1(1/2) + 4(1/4) + 9(1/8) + 16(1/16) + 25(1/32) + \cdots$$

Call this sum E. Formally

$$(1/2)E = 1(1/4) + 4(1/8) + 9(1/16) + 16(1/32) + 25(1/64) + \cdots$$

and subtracting $(1/2)E$ from E we get

$$(1/2)E = 1/2 + 3(1/4) + 5(1/8) + 7(1/16) + 9(1/32) + \cdots$$

Carry out the similar operation to get

$$(1/4)E = 1/2 + 2(1/4) + 2(1/8) + 2(1/16) + 2(1/32) + \cdots$$

Now use the formula of geometric progression to obtain $E = 6$ dollars.]

3.3 LOTTERIES

At the Scarsdale, N.Y., Art Association's annual outdoor show, one sign announced: "Buy a George or Jane Reichart. You may win $50." Taped to each painting was a ticket for the New York State Lottery.

Reader's Digest

Lotteries are games of chance that have existed, some believe, since the Roman emperors Nero and Augustus used them to distribute the slaves. In Europe lotteries were used as early as 1539 to collect state revenue. The first American colonists in Jamestown (1612–1620) were financed largely by lotteries in England; later the colonies conducted their own versions to provide revenues for the new communities. Due to inefficient management and widespread fraud, lottery games were declared illegal in the United States in 1892 by federal law. By 1930 each of the 45 states had passed statutes outlawing lotteries. Lotteries recovered legal status in 1963, first in New Hampshire, followed by New York and New Jersey, and current U.S. laws allow use of television and mail for state lottery development.

A *lottery* is a game of chance with low stakes and potentially high winnings, which account for the widespread appeal of this type of gambling. In its simplest form, a player bets on a number and wins if the state also selects that number. While we usually view a lottery as a game, many applications exist in the real world. For example, insurance is a lottery with the premium of a policy playing the role of the value of a lottery ticket. Compared to the casino games discussed in Section 3.2, the probability of winning a lottery is generally quite low.

Example 3.3.1 Lottery L has 4000 tickets, 1 of which carries a prize of $1000. How appealing is this lottery?

Solution. Let A be the event "$1000 prize"; then $P(A) = 1/4000$. The expected value or *worth* of a ticket is

$$(1000)\left(\frac{1}{4000}\right) + (0)\left(\frac{3999}{4000}\right) = \$0.25$$

We can extend this example to a general lottery with two outcomes and determine the worth of the ticket under indifference, say W; that is, the amount of money which corresponds to indifference between buying the ticket at this price or not buying it. Our discussion is based on the work presented by I. R. Savage (1968) which we highly recommend for further study. In the general case, suppose that a ticket costs \$K. If the event A occurs, the ticket holder receives X_A; otherwise the holder receives $X_{\bar{A}}$. (In Example 3.3.1, A = "\$1000 ticket," X_A = 1000, and $X_{\bar{A}}$ = 0.) Now I would be pleased to purchase a ticket for the amount which is less than the worth of the ticket and would be displeased to purchase it for a larger amount. So if W > K, I am pleased; if W < K, I am displeased because I overpaid; and if W = K, I should be indifferent about this transaction. The worth of the ticket W depends upon the probability of occurrence of event A (more exactly on our assessment of the probability of the event A) and the payoffs X_A and $X_{\bar{A}}$. To find the explicit expression for W we recall the discussion in Section 1.5 of betting an amount of Y dollars on the event A in an elementary gambling situation (EGS) under indifference.

If I pay \$W for a ticket, the change in my assets is

$$X_A - W \quad \text{if A occurs}$$

$$X_{\bar{A}} - W \quad \text{if } \bar{A} \text{ occurs}$$

(3.3.1)

On the other hand, in an EGS the change of assets under the condition or hypothesis of indifference (H) when betting Y dollars on an event A is

$$Y \, \frac{P(\bar{A}|H)}{P(A|H)} \quad \text{if A occurs}$$

$$-Y \quad \text{if } \bar{A} \text{ occurs}$$

(3.3.2)

Since these two betting situations are equivalent, I can equate corresponding parts of Equations (3.3.1) and (3.3.2):

$$X_A - W = Y \, \frac{P(\bar{A}|H)}{P(A|H)}$$

(3.3.3)

$$X_{\bar{A}} - W = -Y$$

(3.3.4)

Substituting Equation (3.3.4) into Equation (3.3.3) and simplifying, we have

$$W = X_A P(A|H) + X_{\bar{A}} P(\bar{A}|H)$$

(3.3.5)

In the lottery L of Example 3.3.1, W = 25¢; we should not be willing to pay more than this for a ticket.

Consider now a lottery L which can result in any one of the three outcomes A, B, and C, which carry the corresponding payoffs X_A, X_B, and X_C.

Suppose our assessments of the probabilities of these particular outcomes are P(A), P(B), and P(C). (Remember that $P(A) + P(B) + P(C) = 1$.) What is the worth of a ticket in lottery L? Consider two related lotteries. Lottery L* is similar to lottery L, except that we know that event A has not occurred; thus the payoffs in lottery L* are X_B, X_C with the probabilities $P(B|\overline{A})$ and $P(C|\overline{A})$, respectively. This is a lottery with two outcomes, and therefore the worth of a ticket is: $W^* = X_B P(B|\overline{A}) + X_C P(C|\overline{A})$. Lottery L** has the payoffs X_A if A occurs and W* if $\overline{A}$ occurs. This is also a lottery with two outcomes, and its expected value is therefore $W^{**} = X_A P(A) + W^* P(\overline{A})$.

A moment of reflection shows that lottery L (with three outcomes A, B, and C) is equivalent to the lottery L** just defined. Indeed, if A occurs both lotteries yield the same reward X_A. If $\overline{A}$ occurs (i.e., if B or C occur) then as far as lottery L is concerned, I have the possibility of rewards X_B and X_C with probabilities $P(B|\overline{A})$ and $P(C|\overline{A})$, and the worth of a ticket for this lottery is $X_B P(B|A) + X_C P(C|A) = W^*$. Lottery L** guarantees payment of W* if $\overline{A}$ occurs (i.e., if $\overline{A}$ occurs the lottery L is equivalent to lottery L**). Thus, lotteries L and L** are equivalent.

Let us now compute the worth of a ticket for lottery L**, which will also be the worth in lottery L. From the explicit expression for W* and the definition of conditional probability, we have

$$W^{**} = X_A P(A) + W^* P(\overline{A}) = X_A P(A) + [X_B P(B|\overline{A}) + X_C P(C|\overline{A})] P(\overline{A})$$
$$= X_A P(A) + X_B P(B \cap \overline{A}) + X_C P(C \cap \overline{A}) \qquad (3.3.6)$$

Now, since B and C are contained in $\overline{A}$, $P(B \cap \overline{A}) = P(B)$ and $P(C \cap \overline{A}) = P(C)$. Thus Equation (3.3.6) becomes

$$W^{**} = X_A P(A) + X_B P(B) + X_C P(C) \qquad (3.3.7)$$

> In general, for a lottery with outcomes $A_1, A_2, \ldots, A_k$ and associated payoffs $X_{A_1}, X_{A_2}, \ldots, X_{A_k}$, the *worth of a ticket* (equivalently, the *expected value of the lottery*) is given by
>
> $$W = X_{A_1} P(A_1) + X_{A_2} P(A_2) + \cdots + X_{A_k} P(A_k)$$

Example 3.3.2

1. Lottery L_1 has 4000 tickets, one carrying a prize of $100 and 30 carrying prizes of $100. What is the worth of a ticket for this lottery? We have

Outcome	Probability	Reward
A = "$1000 prize"	$P(A) = 1/4000$	$X_A = 1000$
B = "$100 prize"	$P(B) = 30/4000$	$X_B = 100$
C = "No prize"	$P(C) = 3969/4000$	$X_C = 0$

Hence, using W_{L_1} for W,

$$W_{L_1} = (1000)\,(1/4000) + (100)\,(30/4000) + (0)\,(3969/4000) = 1$$

The expected value of this lottery is $1.

2. Lottery L_2 also contains 4000 tickets. Twenty tickets carry $100 prizes each, 200 tickets carry $10 prizes each. The expected value is

$$W_{L_2} = (100)\,(20/4000) + (10)\,(200/4000) + (0)\,(3780/4000) = 1$$

3. Lottery L_3 also consists of 4000 tickets. It carries 3 prizes of $1000 and 10 prizes of $100. The expected value for this lottery is

$$W_{L_3} = (1000)\,(3/4000) + (100)\,(10/4000) + (0)\,(3987/4000) = 1$$

The lotteries have the same expected value and should be equivalent; subjectively, however, each one of these lotteries has special appealing features, advantageous in specific situations, which would make it more appropriate than the others. For example, lottery L_3 has the maximum chance of gaining a large $1000 prize, while the maximum chance of winning any prize is in lottery L_2, which contains 220 winning tickets. Lottery L_1 is preferable when the exact amount $100 is required.

Public participation in lotteries is growing nationally as more and more states introduce them. A 1972 survey indicated the following figures for public participation in the four state lotteries.

State	Public participation (%)
New York	59
New Jersey	79
Connecticut	58
Pennsylvania	86

A common characteristic of lotteries is that constant promotion is needed to keep gross revenue at a stable level. The lotteries have shown similar structure across states; a comparison of important aspects of the Ohio and Michigan lotteries is given below (see Rustagi, 1981).

	Michigan	Ohio
Percentage of gross revenue as prizes	45%	45%
Lowest prize	$25	$20
Approximate no. of winners per 1,000,000	4000	5000
Weekly top prize	$200,000	$300,000
Grand prize	$1,000,000	$1,000,000
Commission on ticket sales	5–6%	5–6%

As of February, 1983, state-sponsored lotteries operated in 18 states, including the District of Columbia. The average return per dollar for these lotteries ranged between 44.4 and 55.4 cents with 7 lotteries paying, on average, 50 cents or more per dollar.

Exercises

3.51   A lottery of 1000 tickets offers five prizes of $4000 and eight prizes of $3000. Would you be willing to pay for the ticket:
  a. $20?
  b. $100?
  c. $5?
  Verify your answer by calculations.

3.52   Twenty-five sheets of tickets are sold. Each sheet of tickets contains 10 tickets, and the numbers on each sheet are the same. The prize of $100 is divided proportionally among the 10 tickets with winning numbers (each winning number carries a $10 prize). If I have enough money to buy four tickets, shall I buy them all from the same sheet, each from a different sheet, or two from the same sheet and two from different sheets?

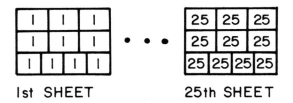

1st SHEET   25th SHEET

3.53   N sheets of tickets are sold. Each sheet of tickets contains n tickets, all with the same number. The prize R dollars is divided proportionally among holders of the winning number; that is, each ticket with a winning number is worth R/n dollars. Suppose we have enough money for m tickets. Shall we buy all m tickets from the same sheet (i.e., bearing the same number) or buy m tickets from m different sheets?

3.54   In the lottery of Exercise 3.53, under what conditions would one procedure be preferable to the other?

3.55   Consider three lotteries: L_1 has 4000 tickes with 1 prize of $1000 and 400 prizes of $10; L_2 has 4000 tickets with 3 prizes of $1000 and 10 prizes of $100; L_3 has 4000 tickets with 20 prizes of $100 and 100 prizes of $1.25.
  a.   Compute the value of each lottery and show that $W_{L_1} > W_{L_2} > W_{L_3}$.

b. Give an example of a situation in which L_2 would be preferable to L_1. Describe a situation in which L_3 would be preferable to L_1.

3.56 The layout of a ticket for the Pennsylvania lottery, known as "Instant Jackpot," is shown below (each play is a combination of 3 figures):

Play 1 $ $ $
Play 2 $ $ $
Play 3 $ $ $

A coin is used to rub the dollar signs from each horizontal row, revealing figures underneath. The possible figures and their corresponding payoffs and probabilities are given below.

Outcome	Payoff ($)	No. of winners per 1,000,000 tickets
A_1 = one cherry on left	1 free ticket plus entry fee	149,992
A_2 = 2 cherries on left	2	83,333
A_3 = 3 cherries	5	16,666
A_4 = 3 oranges	25	1,422
A_5 = 3 stars	50	533
A_6 = 3 bells	100	177
A_7 = 3 sevens	1,000	14.8
A_8 = 3 bars	10,000	3.7

If a ticket costs 50¢, compute the expected value of this version of the Pennsylvania Lottery.

3.57 The New Jersey lottery publishes 1,000,000 tickets per day, each with a five-digit number and costing $1. On a particular day in January, the following table of winning numbers and payoffs was published:

Number	Payoff ($)
69504	10,000
40596	1,000
6950x	225
x9504	225
695xx	25
x950x	25
xx504	25
Qualifier	
12873	100
x2873	50

For the purposes of this exercise, x represents any number between 0 and 9, and a qualifier is considered a regular ticket. Compute the expected value of this version of the New Jersey daily lottery.

4

INFORMATION THEORY

Information for calculating machines, for big computers, for micro-processors, and for us is not what is normally called information. . . . It is actually one of the three fundamental elements of nature, matter, energy, and information, which furnish the materials necessary for our activities and creations.

J. Servan-Screiber

4.1 INFORMATION AS COMMUNICATION

In our discussions the probability of an event measures a degree of belief in its occurrence; that is, probability represents the likelihood of an uncertain event. Let us talk now about the information communicated by the occurrence of an uncertain event. If we observe a rare event, say a snowstorm in Texas in July, then intuitively a large amount of information has been communicated. On the other hand, a commonplace event, such as the sunrise in the morning, conveys hardly any information at all. It is this inverse relationship between probability and information that we will explore in this chapter.

The term *information theory* sounds grandiose enough to solve all the problems of our information-conscious society. The *Random House Dictionary* defines *information* as "knowledge communicated or received concerning a particular fact or circumstance" and *information theory* as "the application of mathematics to language, concepts, processes, and problems in the field of communications." In this chapter we shall develop a quantitative measure of the amount of information contained in the occurrence (or nonoccurrence) of an uncertain event. In general, we will determine the information content of a set $\{E_1, E_2, \ldots, E_n\}$ of mutually exclusive and exhaustive events (see Chapter 1) with corresponding probabilities assessed in a consistent manner.

This definition permits us to address particular applications of communication: language, electronic data processing, and the stock market, for example.

4.2 ENTROPY

Let us begin with a set of n events $\{E_1, E_2, \ldots, E_n\}$ which are exhaustive $[P(E_1 \cup E_2 \cup \cdots \cup E_n) = 1]$ and mutually exclusive $[P(E_i \cap E_j) = 0$ for any $i \neq j]$. We will define a random variable X taking on n different values with probabilities $P(E_1), P(E_2), \ldots, P(E_n)$, respectively. For example, in the toss of a fair coin, E_1 = head, E_2 = tail, and

$$X = \begin{cases} 0 & \text{with probability } P(E_1) = 1/2 \\ 1 & \text{with probability } P(E_2) = 1/2 \end{cases}$$

We are interested in a quantitative expression for the amount of information conveyed by the occurrence of one of the E_i's or equivalently by the random variable X taking on a specific value. As discussed in Section 4.1, this information is related to the uncertainty of the set of events before the occurrence of E_i particularly, and we will use the terms *amount of information* and *amount of uncertainty* interchangeably.

The relationship of information theory to statistical thermodynamics (Brillouin, 1956) has caused the term *entropy* to denote the amount of information or uncertainty in a collection of mutually exclusive events; we shall use H(X) to denote the entropy in the random variable X. Let us think for a moment about some of the properties we would like (intuitively) for this function H(X) to have. Consider first an experiment with k equally likely outcomes. In this case, the uncertainty of the experiment is dependent only on the number k; let's call it $f(k)$. If, for example, $k = 1$, then the experiment has a certain outcome, and the uncertainty or information is zero. The larger the value of k, the more difficult the task of predicting the outcome; that is, the information is an increasing function of k. Intuitively, then, we would like to have

$$f(1) = 0 \qquad\qquad (4.2.1)$$

and

$$f(k) \text{ increases with k} \qquad\qquad (4.2.2)$$

Consider now two independent experiments, that is, two independent random variables X and Y (see Section 1.2 for a discussion of independence). If X and Y have k and ℓ equally likely outcomes, respectively, then the joint occurrence of both X and Y, denoted by $X \otimes Y$, has k times ℓ equally likely outcomes. For example, let X be 0 or 1, according to whether the outcome of a toss of a fair coin is tails or heads, respectively, and let Y denote the face showing on the roll of a fair die. Here k = 2, ℓ = 6, and the 12 equally likely outcomes of X ⊗ Y are the following:

Outcome	Value of X	Value of Y
1	0	1
2	0	2
3	0	3
4	0	4
5	0	5
6	0	6
7	1	1
8	1	2
9	1	3
10	1	4
11	1	5
12	1	6

The amount of information in the compound experiment "tossing a fair coin and rolling a fair die" should be equal to the sum of the information contained in the components: if you know the joint outcome, then you know the component outcomes, and vice versa. Thus our information measure should satisfy

$$f(k\ell) = f(k) + f(\ell) \tag{4.2.3}$$

for any integers k and ℓ.

Now, if we muse for a moment on the intuitively appealing relations of Equations (4.2.1), (4.2.2), and (4.2.3), we will realize that they are all satisifed by the *logarithmic* function $f(k) = \log(k)$; furthermore, this is the *only* (continuous) function which has these three properties. For any choice of the base, recall the properties of logarithms: $\log 1 = 0$; larger numbers have larger logarithms (e.g., $\log_{10} 10 = 1$ and $\log_{10} 100 = 2$); and for positive numbers a and b, $\log(ab) = \log(a) + \log(b)$ [e.g., $\log_{10}(10 \cdot 100) = \log_{10} 1000 = 3 = \log_{10}(10) + \log_{10}(100)$].

To specify the base for our logarithmic entropy function, let us take as our unit of entropy the amount of information contained in an experiment with two equally likely outcomes. Examples include the toss of a fair coin, a true/false question for which each answer is equally likely, a switch in a computer circuit which is equally likely to be on or off in any given second. From the computer analogy, we take the term *bit* to denote this unit of entropy. Then, clearly the base of our logarithmic function must be 2, since $\log(2) = 1$ bit. The entropy of an experiment with 10 equally likely outcomes is then $\log_2(10) = 3.32$ bits. Thus we have developed the following definition.

The *entropy* of an experiment with k equally likely
outcomes is given by: $f(k) = \log_2(k)$ bits. (4.2.4)

In the remainder of this chapter we will denote $\log_2(k)$ by $\log(k)$.

As we have seen repeatedly in this volume, many real-world phenomena
do not accommodate themselves to our notion of equally likely outcomes, so
let us determine the entropy of an experiment with k possible outcomes, not
necessarily equally probable. Consider the decomposition

$$\log k = -\log \frac{1}{k} = -\underbrace{\frac{1}{k} \log \frac{1}{k} - \frac{1}{k} \log \frac{1}{k} - \cdots - \frac{1}{k} \log \frac{1}{k}}_{\text{k terms}}$$ (4.2.5)

as the entropy in an experiment with k equally likely outcomes, each with
probability $1/k$ and contribution $(-1/k) \log (1/k)$ to the total entropy of the
experiment. Extending this idea to the case of k outcomes $E_1, \ldots, E_k$, with
probabilities $P(E_1), \ldots, P(E_k)$, respectively, so that $P(E_1) + \cdots + P(E_k) = 1$,
the contribution of outcome E_i to total entropy is

$$-P(E_i) \log P(E_i)$$

and we arrive at the following extension of Definition (4.2.4).

Let X be a random variable taking on k discrete values with probabilities $p_1, p_2, \ldots, p_k$, respectively; then the entropy of the random
variable X is given by

$$H(X) = -p_1 \log p_1 - p_2 \log p_2 - \cdots - p_k \log p_k$$

$$= -\sum_{i=1}^{k} p_i \log p_i$$ (4.2.6)

Note that the entropy of a random experiment depends only on the probabilistic
structure of the outcomes and not on the actual values of the random variable.

Example 4.2.1 Let E_1, E_2, E_3 be mutually exclusive events with $P(E_1) = 1/2$
and $P(E_2) = P(E_3) = 1/4$. Then the entropy of this experiment is

$$H(X) = -\frac{1}{2} \log \frac{1}{2} - \frac{1}{4} \log \frac{1}{4} - \frac{1}{4} \log \frac{1}{4}$$

$$= \frac{1}{2} + \frac{1}{2} + \frac{1}{2}$$

$$= 1.5 \text{ bits}$$

Had the three outcomes been equally probable, the entropy would have been log 3 $\approx$ 1.57 bits. In general, the entropy for any collection of outcomes is largest when the outcomes are equally probable. This is intuitively plausible, since the most uncertain case is when no one outcome is preferable to any other.

Example 4.2.2 This example presents an important aspect of entropy as defined in Equation (4.2.6). Consider the random outcome X of tomorrow's weather where three possible outcomes are

Outcome	Probability
E_1 = rain	p_1 = 0.3
E_2 = snow	p_2 = 0.1
E_3 = clear skies	p_3 = 0.6

Let us combine E_1 and E_2 into a single (coarse) outcome $B = E_1 \cup E_2$ = precipitation, and consider the resulting experiment Y having two outcomes B and E_3 with probabilities 0.4, and 0.6, respectively. Intuitively, H(Y) < H(X) and in particular

$$H(X) = -0.3 \log 0.3 - 0.1 \log 0.1 - 0.6 \log 0.6$$

$$= (-0.4 \log 0.4 - 0.6 \log 0.6) + 0.4 \left(-\frac{0.3}{0.4} \log \frac{0.3}{0.4} - \frac{0.1}{0.4} \log \frac{0.1}{0.4} \right)$$

$$= H(Y) + 0.4\, H(Z)$$

where the first term represents the entropy of the coarse experiment Y and Z represents the two-outcome experiment with probability structure 0.3/0.4, 0.1/0.4. [Note that H(Z) cannot be negative.] In general, for an experiment with three outcomes and probabilities p_1, p_2, p_3,

$$H(p_1, p_2, p_3) = H(p_1 + p_2, p_3) + (p_1 + p_2) H\left(\frac{p_1}{p_1 + p_2}, \frac{p_2}{p_1 + p_2} \right) \quad (4.2.7)$$

Indeed X and Y are equivalent if E_3 occurs; if E_1 or E_2 occurs, X yields more information than Y and thus H(X) > H(Y). When Y indicates that either E_1 or E_2 has occurred, we must perform a second experiment Z with two (conditional) outcomes $E_1|(E_1 \cup E_2)$ and $E_1|(E_1 \cup E_2)$ and corresponding probabilities $p_1/(p_1 + p_2)$ and $p_2/(p_1 + p_2)$, respectively, to determine which of E_1 or E_2 actually occurred.

More analytically, Equation (4.2.7) can be derived as follows. First,

$$H(p_1, p_2, p_3) = -p_1 \log p_1 - p_2 \log p_2 - p_3 \log p_3 \quad (4.2.8)$$

$$H(p_1 + p_2, p_3) = -(p_1 + p_2) \log (p_1 + p_2) - p_3 \log p_3 \quad (4.2.9)$$

$$H\left(\frac{p_1}{p_1+p_2},\ \frac{p_2}{p_1+p_2}\right) = -\ \frac{p_1}{p_1+p_2}\ \log\ \frac{p_1}{p_1+p_2}\ -\ \frac{p_2}{p_1+p_2}\ \log\ \frac{p_2}{p_1+p_2}$$

$$(4.2.10)$$

Multiplying Equation (4.2.10) by (p_1+p_2) and adding the result to Equation (4.2.9), we have

$$H(p_1+p_2,p_3)+(p_1+p_2)H\left(\frac{p_1}{p_1+p_2},\ \frac{p_2}{p_1+p_2}\right)$$

$$=\ -p_1\ \log(p_1+p_2)-p_2\ \log(p_1+p_2)-p_3\ \log\ p_3$$

$$-p_1\ [\log\ p_1-\log(p_1+p_2)]-p_2\ [\log\ p_2-\log(p_1+p_2)] \qquad (4.2.11)$$

since $\log(a/b) = \log a - \log b$. Elimination of identical terms reduces Equation (4.2.11) to $H(p_1,p_2,p_3)$.

Another approach to the entropy of an experiment defined by the random variable X is the average uncertainty of the experiment. In Equation (4.2.6), if the quantity $-\log p_i$ represents the information in a particular outcome E_i with probability p_i, then H(X) is just the expected value of the informations of the particular outcomes. Consider the following example.

Example 4.2.3 Recall the experiment of Example 4.2.1. If E_1 occurs, it contributes $-\log(1/2) = \log 2 = 1$ bit of information. If E_2 occurs, it contributes $-\log(1/4) = \log 4 = 2$ bits of information, the same as the potential contribution of E_3. On the average, the whole experiment supplies $(1/2)(1) + (1/4)(2) + (1/4)(2) = 1.5$ bits of information.

Calculation of entropies is facilitated by Table B.4 (Appendix B) of values of $-p\ \log_2 p$ for values of p from 0.001 to 0.999. For example, the entropy of an experiment with probabilities 0.4 and 0.6 is evaluated from the table as $0.5288 + 0.4422 = 0.971$ bit (0.029 bit less than the information conveyed by a two-outcome experiment with equal probabilities).

Exercises

4.1 Prove that any reasonable measure of information must satisfy:

$$H(p_1,p_2,p_3,p_4) = H(p_1+p_2,p_3+p_4)$$

$$+\ (p_1+p_2)\ H\left(\frac{p_1}{p_1+p_2},\ \frac{p_2}{p_1+p_2}\right)$$

$$+\ (p_3+p_4)\ H\left(\frac{p_3}{p_3+p_4},\ \frac{p_4}{p_3+p_4}\right)$$

and that our definition satisfies this relationship.

4.2   Two classes contain 20 students each. The first class consists of 10 fresh-men, 5 sophomores, and 5 juniors; the second consists of 8 freshmen, 8 sophomores, and 4 juniors. A student is chosen randomly from each of the classes. Observing the level of a student from which class is more informative?

4.3   How much information is contained in the experiment of asking a "person on the street" whether today is his or her birthday? Interpret your result.

4.4   You asked someone to guess a number between 1 and 10. What is the information contained in finding out what number was guessed?

4.5   In computer technology, signals of 1 or 0 (on or off) are important means of communication. Suppose we wish to send one such signal, and to increase the chances that our intended message is received we send three signals. To communicate a 1 we send three 1s, and to communicate a 0 we send three 0s. At the receiving end, the possible four groups of signals are those involving three 1s, two 1s and one 0, two 0s and one 1, or three 0s. Using a majority rule, the first two cases (that is, three 1s or two 1s and one 0) are taken to mean 1, and the last two are taken to mean 0. Suppose we need to send a 1, and the probability of any signal failing (transmitting an erroneous message) is 0.1. What is the entropy in the experiment consisting of observing the message at the receiving end? [Hint: The probability of receiving "1 1 0" when "1 1 1" is sent is $(0.9)^2 (0.1)$, and so on.]

4.3   MUTUAL INFORMATION

Our discussion in Section 4.2 was of the amount of information in a random experiment with a finite number of possible outcomes. In many real-world situations we are faced with determining the amount of information in an experiment contained in another closely related experiment. For example, we hear a forecast (X) about tomorrow's weather (Y) or we perform a seismic test (X) to determine whether oil is present (Y).

Let us take a random experiment X with outcomes x_1, x_2, and x_3 and corresponding probabilities $p(x_1)$, $p(x_2)$, and $p(x_3)$ and another random experiment Y with two outcomes y_1 and y_2 and probabilities $p(y_1)$ and $p(y_2)$. For example, we might take

$$Y = \begin{cases} 1 & \text{if oil is present} \\ 0 & \text{if no oil is present} \end{cases}$$

$$X = \begin{cases} -1 & \text{if seismic test is negative} \\ 0 & \text{if seismic test is inconclusive} \\ 1 & \text{if seismic test is positive} \end{cases}$$

The compound experiment $X \otimes Y$ has 6 outcomes $(x_1, y_1), \ldots, (x_3, y_2)$ with probabilities given by

$$p(x_i, y_j) = p(x_i)\, p(y_j | x_i)$$
$$= p(y_j)\, p(x_i | y_j) \qquad i = 1, 2, 3; \; j = 1, 2$$

where, e.g., $p(y_j | x_i)$ is the conditional probability of outcome y_j in experiment Y given that x_i is the outcome of experiment X.

If X and Y are independent random variables, then $p(x_i, y_j) = p(x_i)\, p(y_j)$ ($i = 1, 2, 3; j = 1, 2$). In this case the entropy of the compound experiment is the sum of the component entropies. We see this analytically as follows.

$$
\begin{aligned}
H(X \otimes Y) &= -p(x_1, y_1) \log p(x_1, y_1) - p(x_1, y_2) \log p(x_1, y_2) \\
& \quad - \cdots - p(x_3, y_2) \log p(x_3, y_2) \\
&= -p(x_1) p(y_1) [\log p(x_1) + \log p(y_1)] \\
& \quad -p(x_1) p(y_2) [\log p(x_1) + \log p(y_2)] \\
& \quad - \cdots - p(x_3) p(y_2) [\log p(x_3) + \log p(y_2)] \\
&= -p(x_1) p(y_1) \log p(x_1) - p(x_1) p(y_1) \log p(y_1) \\
& \quad -p(x_1) p(y_1) \log p(x_1) - p(x_1) p(y_1) \log p(y_2) \\
& \quad - \cdots \\
& \quad -p(x_3) p(y_2) \log p(x_3) - p(x_3) p(y_2) \log p(y_2) \\
&= -p(x_1) \log p(x_1) [p(y_1) + p(y_2)] \\
& \quad -p(x_2) \log p(x_2) [p(y_1) + p(y_2)] \\
& \quad - \cdots - p(y_2) \log p(y_2) [p(x_1) + p(x_2) + p(x_3)] \\
&= H(X) + H(Y)
\end{aligned}
$$

since $p(y_1) + p(y_2) = p(x_1) + p(x_2) + p(x_3) = 1$.

If a compound experiment has independent components, information in the compound experiment is the sum of informations in the components:

$$H(X \otimes Y) = H(X) + H(Y) \qquad (4.3.1)$$

If X and Y are not independent, the conditional relation $p(x, y) = p(x)\, p(y|x)$ will yield

$$
\begin{aligned}
H(X \otimes Y) = H(X) \\
+ p(x_1)[-p(y_1 | x_1) \log p(y_1 | x_1) - p(y_2 | x_1) \log p(y_2 | x_1)] \\
+ p(x_2)[-p(y_1 | x_2) \log p(y_1 | x_2) - p(y_2 | x_2) \log p(y_2 | x_2)] \\
+ p(x_3)[-p(y_1 | x_3) \log p(y_1 | x_3) - p(y_2 | x_3) \log p(y_2 | x_3)]
\end{aligned}
$$

$$(4.3.2)$$

The expression $-p(y_1|x_1)\log p(y_1|x_1) - p(y_2|x_1)\log p(y_2|x_1)$, now called $H(Y|x_1)$, is the uncertainty contained in experiment Y given that outcome x_1 of experiment X occurred. Similarly, the factor multiplying $p(x_2)$ is $H(Y|x_2)$ and the factor multiplying $p(x_3)$ is $H(Y|x_3)$. Thus the expression following $H(X)$ can be written

$$p(x_1)H(Y|x_1) + p(x_2)H(Y|x_2) + p(x_3)H(Y|x_3) = H(Y|X)$$

the average uncertainty about Y after (or given that) an outcome of X occurs. $H(Y|X)$ is called the *conditional entropy of Y given X*. If the experiments X and Y are independent, then $H(Y|x_1) = H(Y|x_2) = H(Y|x_3) = H(Y)$ and thus also $H(Y|X) = H(Y)$ since $p(x_1) + p(x_2) + p(x_3) = 1$.

Returning to Equation (4.3.2) for $H(X \otimes Y)$ and using the decomposition $p(x, y) = p(y)p(x|y)$, we now obtain

$$\begin{aligned} H(X \otimes Y) = \ & H(Y) + \big\{ p(y_1)[-p(x_1|y_1)\log p(x_1|y_1) \\ & - p(x_2|y_1)\log p(x_2|y_1) - p(x_3|y_1)\log p(x_3|y_1)] \\ & + p(y_2)[-p(x_1|y_2)\log p(x_1|y_2) - p(x_2|y_2)\log p(x_2|y_2) \\ & - p(x_3|y_2)\log p(x_3|y_2)] \big\} \end{aligned}$$

The factor multiplying $p(y_1)$ is $H(X|y_1)$, the uncertainty contained in the experiment X after the outcome y_1 of Y occurred, and the whole expression contained in braces to the right of $H(Y)$ is thus

$$p(y_1)H(X|y_1) + p(y_2)H(X|y_2) = H(X|Y)$$

the average uncertainty (or information) contained in X given that an outcome of Y occurred. If Y and X are independent, then $H(X|y_1) = H(X|y_2) = H(X|y_3) = H(X|Y) = H(X)$, i.e., the occurrence of Y does not affect the uncertainty (or information) contained in X. Note that in general $H(X|Y) \leqslant H(X)$, but for a particular value of Y, y_1 say, we may have $H(X|y_1) > H(X)$ (see Exercise 4.6, reversing the roles of X and Y).

Suppose, however, the outcome of Y determines the outcome of X (completely), as might be the case for our example where Y denotes presence of oil and X denotes the test results. For example, if $p(x_1|y_1) = 1$, $p(x_2|y_1) = 0$ and $p(x_3|y_1) = 0$, $p(x_1|y_2) = 0$, $p(x_2|y_2) = 1$ and $p(x_3|y_2) = 0$, then $H(X|Y)$ is zero. [Formally we obtain here quantities of the form $-1\log(1) - 0\log(0)$; here the first summand is zero because $\log 1 = 0$, while the second is zero by continuity of the log function (Table B.4). Recall that the information in an experiment with a sure outcome is zero.]

Consider now the difference $H(X) - H(X|Y)$. $H(X)$ is the (average) uncertainty contained in experiment X and $H(X|Y)$ is the (average) uncertainty contained in X after an outcome Y is known. If X and Y are dependent, intuitively we know that $H(X|Y)$ is less than $H(X)$, since knowledge of Y reduces the uncertainty about X, and the difference $H(X) - H(X|Y) = I(X; Y)$ is the amount of information about X contained in Y. If $H(X|Y) = H(X)$, then $I(X; Y) = 0$ and there is *no* information about X contained in Y; if however $H(X|Y) = 0$ (i.e., Y completely determines X) then the information $I(X; Y)$ is maximal and is equal to $H(X)$, the information about X contained in X! Similarly, we may consider the difference $H(Y) - H(Y|X) = I(Y; X)$, which is by analogy the amount of information about Y contained in X. In Exercise 4.7 you are asked to show that $I(X; Y) \equiv H(X) - H(X|Y) = H(Y) - H(Y|X) \equiv I(Y; X)$, i.e., our definition yields that the amount of information about X contained in Y is equal to the amount of information about Y contained in X and this information is called the *(average) mutual information between X and Y*.

We have four equivalent formulas for the mutual information.

The (average) *mutual information* between two experiments X and Y is given by the following equivalent formulas:

$$I(X; Y) = H(X) - H(X|Y)$$

$$= - \sum_i p(x_i) \log p(x_i)$$

$$+ \sum_j p(y_i) \left[\sum_i p(x_i | y_j) \log p(x_i | y_j) \right] \qquad (4.3.3)$$

$$I(X; Y) = H(Y) - H(Y|X)$$

$$= - \sum_i p(y_j) \log p(y_j)$$

$$+ \sum_i p(x_i) \left[\sum_j p(y_j | x_i) \log p(y_j | x_i) \right] \qquad (4.3.4)$$

$$I(X; Y) = \sum_j \sum_i p(x_i, y_j) \log \frac{p(x_i, y_j)}{p(x_i) p(y_j)} \qquad (4.3.5)$$

$$I(X; Y) = H(X) + H(Y) - H(X \otimes Y) \qquad (4.3.6)$$

Equation (4.3.5) is especially useful when tables of $-p \log p$ are not available. Notice the similarity between Equation (4.3.6) and Equation (1.4.2) for the probability of the union of two events. You are asked to investigate these relationships further in the exercises following this section.

Example 4.3.1 Consider the following conditional distribution table:

		X	
$p(x|y)$		x_1	x_2
	y_1	0	1
Y	y_2	1	0
	y_3	1/2	1/2

and these unconditional probabilities for Y: $p(y_1) = p(y_2) = p(y_3) = 1/3$. The mutual information can be computed using any of the four equivalent formulas, and we illustrate each of these.

Equation (4.3.3). $I(X; Y) = H(X) - H(X|Y)$.

$$p(x_1) = p(y_1)p(x_1|y_1) + p(y_2)p(x_1|y_2) + p(y_3)p(x_1|y_3)$$
$$= (1/3)(0) + (1/3)(1) + (1/3)(1/2) = 1/2$$

Hence, $p(x_2) = 1 - p(x_1) = 1/2$, and thus $H(X) = -1/2 \log (1/2) - 1/2 \log (1/2) = 1$ bit. Now,

$$H(X|Y) = p(y_1)[-p(x_1|y_1)\log p(x_1|y_1) - p(x_2|y_1)\log p(x_2|y_1)] + \cdots$$
$$+ p(y_3)[-p(x_1|y_3)\log p(x_1|y_3) - p(x_2|y_3)\log p(x_2|y_3)]$$

$$= \frac{1}{3}(-0 \log 0 - 1 \log 1) + \frac{1}{3}(-1 \log 1 - 0 \log 0)$$

$$+ \frac{1}{3}\left(-\frac{1}{2}\log\frac{1}{2} - \frac{1}{2}\log\frac{1}{2}\right)$$

$$= \frac{1}{3} \text{ bit}$$

Hence $I(X; Y) = 1 - 1/3 = $ *2/3 bit*. [This is the most convenient formula in this case, since the table of $p(x|y)$ is given.]

Equation (4.3.4). $I(X; Y) = H(Y) - H(Y|X)$. In our case

$$H(Y) = -\frac{1}{3}\log\frac{1}{3} - \frac{1}{3}\log\frac{1}{3} - \frac{1}{3}\log\frac{1}{3}$$

$$= \log(3) \text{ bits}$$

$$H(Y|X) = p(x_1)[-p(y_1|x_1)\log p(y_1|x_1) - \cdots - p(y_3|x_1)\log p(y_3|x_1)]$$
$$+ p(x_2)[-p(y_1|x_2)\log p(y_1|x_2) - \cdots - p(y_3|x_2)\log p(y_3|x_2)]$$

Now

$$p(y_1|x_1) = \frac{p(y_1, x_1)}{p(x_1)} = \frac{p(x_1|y_1)p(y_1)}{p(x_1)} = 0$$

$$p(y_2|x_1) = \frac{2}{3}$$

$$p(y_3|x_1) = \frac{1}{3}$$

$$p(y_1|x_2) = \frac{(1)(1/3)}{1/2} = \frac{2}{3}$$

$$p(y_2|x_2) = 0$$

$$p(y_3|x_2) = \frac{1}{3}$$

Thus

$$H(Y|X) = \frac{1}{2}\left(-\frac{2}{3}\log\frac{2}{3} - \frac{1}{3}\log\frac{1}{3}\right) + \frac{1}{2}\left(-\frac{2}{3}\log\frac{2}{3} - \frac{1}{3}\log\frac{1}{3}\right)$$

$$= -\frac{2}{3}\log\frac{2}{3} - \frac{1}{3}\log\frac{1}{3}$$

Hence

$$I(X;Y) = \log 3 + \frac{2}{3}\log\frac{2}{3} + \frac{1}{3}\log\frac{1}{3}$$

$$= \frac{2}{3}\log 3 + \frac{2}{3}\log\frac{2}{3}$$

$$= \frac{2}{3} \text{ bit}$$

Equation (4.3.5).

$$I(X;Y) = \sum_j \sum_i p(x_i, y_j) \log \frac{p(x_i, y_j)}{p(x_i)p(y_j)}$$

$$I(X;Y) = 0 \log \frac{0}{(1/2)(1/3)} + \frac{1}{3}\log\frac{1/3}{(1/2)(1/3)} + \frac{1}{3}\log\frac{1/3}{(1/2)(1/3)}$$

$$+ 0 \log \frac{0}{(1/2)(1/3)} + \frac{1}{6}\log\frac{1/6}{(1/2)(1/3)} + \frac{1}{6}\log\frac{1/6}{(1/2)(1/3)}$$

$$= \frac{1}{3} + \frac{1}{3} = \frac{2}{3} \text{ bit}$$

(Log 1 = 0 and 0 log 0 = 0.)

Equation (4.3.6). $I(X;Y) = H(X) + H(Y) - H(X \otimes Y)$. In this case we compute the joint probabilities $p(x_1, y_1) = 0$, $p(x_1, y_2) = 1/3, \ldots, p(x_2, y_3) = 1/6$, and obtain

$$I(X;Y) = 1 + \log 3 + \frac{2}{3}\log\frac{1}{3} + \frac{1}{3}\log\frac{1}{6}$$

$$= 1 + \log 3 - \frac{2}{3}\log 3 - \frac{1}{3}\log 2 - \frac{1}{3}\log 3$$

$$= \frac{2}{3} \text{ bit}$$

Exercises

4.6   A box contains four black and six white balls. Experiments X and Y
       are two successive drawings of a ball from the box. Let the outcomes
       be x_1 = a black ball on the first drawing and x_2 = a white ball on the
       first drawing; define similarly for the experiment Y. Compute $H(X)$,
       $H(Y)$, $H(Y|X)$, $H(Y|x_1)$ and $H(Y|x_2)$. [Note that $H(Y|X) < H(Y)$,
       but $H(Y|x_2) > H(Y)$.]

4.7   Show that $H(X) - H(X|Y) = H(Y) - H(Y|X)$. [Hint: Use the fact that
       $\sum_j p(x_i, y_j) = p(x_i)$.]

4.8   In New Jersey the probability of a snowy day in January is 0.4. One
       weather reporter is correct in 3/5 of the cases when she predicts snow
       and in 4/5 of the cases when she predicts fair weather. What is the
       amount of information contained in her forecast? Comment on her
       efficiency; i.e., how often, on the average, are her forecasts correct?

4.9   The forecaster of Exercise 4.8 is trying to improve her record in
       February when the probability of a snowy day is 0.8. She is correct in
       9/10 of the cases when snow is predicted and in only 1/2 of the cases
       when fair weather is predicted. Compute the information about the
       weather in her February forecast. (Note that the worst possible case
       results from a forecaster being correct in exactly 1/2 of the forecasts,
       since such a record yields maximum possible uncertainty.)

4.10   Out of every 100 sites selected for testing for oil, two actually contain oil
       If oil is present, a soil test is positive; if no oil is present, then the chances
       are 50:50 that the test will have a positive result. What is the amount of
       information about the state of the site contained in the test? Comment
       on the magnitude of your answer, compared to the maximum informa-
       tion obtained in an experiment with two outcomes.

4.11   For the following joint table, compute $H(Y|X)$ and the mutual informa-
       tion $I(X; Y)$ using the formula $H(Y) - H(Y|X)$. Also, verify that
       $H(X \otimes Y) < H(X) + H(Y)$.

		X	
$p(x, y)$	x_1	x_2	x_3
y_1	0.10	0.15	0.05
Y y_2	0.05	0.03	0.02
y_3	0.30	0.20	0.10

4.12   In a production plant, 70 percent of the parts are lightweight. Among
       these lightweight parts, 80 percent are round. The total percentage of
       round parts is 64 percent. Find the information about the weight of
       the part contained in its shape.

4.13   A random variable X takes two values x_1 and x_2 with equal probability.
       A random variable Y takes two values y_1 and y_2 with conditional proba-
       bilities $p(y_1|x_1) = p(y_2|x_2) = 0.9$, and $p(y_2|x_1) = p(y_1|x_2) = 0.1$. Verify

that $I(X; Y) = 0.532$, using two different calculation formulas. (Hence, a channel with only 10 percent distortion reduces the mutual information by almost 50 percent, from 1 to 0.532!)

4.14 Compute $I(X; Y)$ for the following joint probability tables:

$p(x, y)$	x_1	x_2	x_3
y_1	0.1	0	0
y_2	0	0	0.5
y_3	0	0.4	0

$p(x, y)$	x_1	x_2	x_3
y_1	0	0	0.5
y_2	0	0.1	0
y_3	0.4	0	0

$p(x, y)$	x_1	x_2
y_1	0.1	0
y_2	0.5	0
y_3	0	0.4

4.15 In the computer transmission application of Exercise 4.5, let X denote the signal we wish to send with $P(X = 0) = 0.5 = P(X = 1)$. Let Y denote the three signals received, where each signal has failure rate 0.1.
   a. Compute $I(X; Y)$ for this channel.
   b. What is the maximum possible value for $I(X; Y)$ for a channel with with two input and three output signals?

4.16 Consider the quantity $I(X; Z|Y) = H(X|Y) - H(X|Y \otimes Z)$.
   a. Give a verbal interpretation of this quantity.
   b. Show that $I(X; Z|Y) \geqslant 0$.
   c. What is the maximum possible value of $I(X; Z|Y)$ and what is its practical meaning?
   d. What is the practical meaning of $I(X; Z|Y) = 0$?
   e. Show that $I(X; Z|Y) = I(Z; X|Y)$.
   f. Show that $I(X; Y|Z) = I(X; Y) + I(X; Z|Y)$.
   g. Show that $I(X; Y|Z) + I(Z; Y) = I(Y; X|Z) + I(Z; X)$.
   h. What is the relation between X, Y, and Z which will give $I(X; Y|Z) = I(X; Y)$?

4.4 APPLICATIONS OF INFORMATION

In this section we illustrate the wide range of problems that can be addressed by information theoretic concepts.

Example 4.4.1 Consider the experiment X of guessing a particular integer between 1 and 10; the concept of entropy allows us to put an upper bound on the number of yes or no questions necessary to guess the given integer. Here, $H(X) = \log_2(10) = 3.32$ bits, and the maximum amount of information in one answer to a yes or no question is 1 bit (for the case in which the answers are equally probable). Hence, four questions, yielding four bits of information, should be enough to guess the given integer. A suitable sequence of questions might begin: (1) Is the number less than or equal to 5? The answer to question (1) will reduce the uncertainty by 1 bit. If that answer is yes, ask: (2) Is the number less than or equal to 2? If the answer to question (1) is no, ask: (2) Is the number less than or equal to 7? Clearly, four such questions are sufficient to guess any integer up to $2^4 = 16$.

Example 4.4.2 In a famous puzzle you are given a balance and asked to sort out from nine balls in three weighings the one ball which has a different weight from the rest and determine whether it is heavier or lighter. Since the odd ball is equally likely to be any of the nine and we do not know whether it is lighter or heavier, there are $(2)(9) = 18$ possible outcomes (first ball is heavier, first ball is lighter, etc.). Thus the information in this experiment is $\log_2 18 = 4.16$ bits.

Each weighing can have three results: either of the pans could be heavier or the pans could balance; so each weighing yields a maximum of 3 bits of information. Then three weighings yield log 27 bits, sufficient to find the odd ball. The maximum gain of information is obtained if the first move is to divide the balls into three equal piles, A, B, and C, and to weigh two of them, say A and B. Then the probability that the scales balance is the probability that the odd ball is in pile C, 1/3, which is also the probability that either side is heavier. So we get the maximum possible gain in entropy, $-\log(3) = -1.58$. We detail this procedure as follows.

Step 1.   Put balls 1, 2, and 3 on the left pan and 4, 5, and 6 on the right pan; leave 7, 8, and 9 unweighed.

Step 2.

a.   If the scales balance in step 1, the defective ball is 7, 8, or 9, and we have reduced the uncertainty from log 18 to log 18 − log 3 = log 6 bits. Now put one ball (7, 8, or 9) in each pan and leave one unweighed. If the pans balance, the defective ball is the unweighed one, which can then be compared to another to determine whether it is lighter or heavier. The other two possibilities are analogous.

b.   Suppose the scales tilt in step 1, without loss of generality as in Figure 4.1. Then one of the side 1, 2, 3 is lighter or one of the side 4, 5, 6 is heavier,

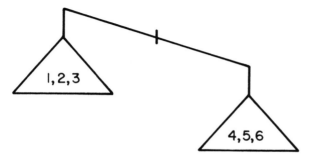

Figure 4.1   Balance and balls configuration for Example 4.4.2.

and the uncertainty is reduced from log 18 to log 6. To create equally likely outcomes for the second weighing, remove 1 and 4 from the balance, interchange 2 and 5, and leave 3 and 6 unchanged.

Step 3. If step 2 yields a balance, then ball 1 is lighter or ball 4 is heavier; if the imbalance is in the same direction as step 1, then either 3 is lighter or 6 is heavier; if the imbalance is in the opposite direction, either 2 is lighter or 5 is heavier. In any of these cases, the entropy is reduced from log 6 to log 2 and the third weighing eliminates this uncertainty. A comparison of the suspicious ball with another ball can yield only two possible outcomes:

a. Balance, indicating that the suspicious ball is not the odd one and the other suspicious ball is
b. Imbalance in one direction only, depending on whether the suspicious ball is suspected to be lighter or heavier

You are asked to investigate other properties of the solution in the exercises. These are very instructive and we recommend them highly.

Example 4.4.3   In this example we consider the application of information theory to linguistics, by approximating the information contained in a single letter of an English text. If all letters and the space character appeared in the text with equal frequency, the entropy of English text would be $\log 27 = 4.75$ bits/symbol. This value is called the zero estimate of the entropy, we write $H_0 = 4.75$. However, in a typical text, letters do not appear with equal frequency; Table 4.1 gives an empirical frequency distribution based on a large amount of experimental data. Thus the actual entropy of English text is decreased (why?) from H_0 to

$$H_1 = \sum_{i=1}^{27} p_i \log p_i = 4.03 \text{ bits/symbol}$$

We are still overestimating the value of uncertainty, since we are assuming that the letters occur independently of each other. However, it is clear that the letters are not independent, since the probability of a letter being h given that the previous letter is t is much larger than the probability of an h given that the previous letter was z; that is, $p(h|t) > p(h|z)$. Also, $p(x|x)$ is usually very small, unless you happen to work for Exxon! So we must find probabilities $p(i|j)$ for all 27 characters of the alphabet (including the space) and then use the formula for conditional entropy from Section 4.2

$$H_2 = -\sum_{i=1}^{27} p(i) \left[\sum_{j=1}^{27} p(j|i) \log p(j|i) \right]$$
$$= -\sum_{i,j} p(i,j) \log (p(j|i))$$

Table 4.1   Empirical Probabilities of English Symbols

Symbol	Empirical probability (p_i)	Symbol	Empirical probability (p_i)
A	0.0642	N	0.0574
B	0.0127	O	0.0632
C	0.0218	P	0.0152
D	0.0317	Q	0.0008
E	0.1031	R	0.0484
F	0.0208	S	0.0514
G	0.0152	T	0.0796
H	0.0467	U	0.0228
I	0.0575	V	0.0083
J	0.0008	W	0.0175
K	0.0049	X	0.0013
L	0.0321	Y	0.0164
M	0.0198	Z	0.0005
		space	0.1859

Source: Abramson, 1963.

Calculations show that H_2 for an English text is 3.32 bits/symbol. Moreover, the dependency between the letters in an English text extends beyond the first preceeding letter (for example, $p(n|oi) > p(n|pl)$) and a better approximation would consider the (average) conditional entropy given two preceding letters, and so on. Extensive empirical investigations indicate that $H_3 = 3.10$ bits/symbol, $H_5 = 2.1$ bits/symbol, and $H_8 = 1.9$ bits/symbol. These investigations indicate that there is no decrease in approximations after H_{30}, so we use this value as a reasonable approximation to the uncertainty of English text. Even using modern computers, we cannot evaluate the exact value of H_{30} (we would need to take into account all n-tuples of 30 letters) but by iterative methods we estimate that H_{30} is approximately 1.3 bits/symbol. We see that the redundancy of an English text is thus $1 - (1.3/4.75) = 0.7$; that is, the letters of the text are so dependent that 70 percent of the structure of the text is devoted to satisfying specific linguistic and grammatical constraints and only 30 percent is "free." In other words, if in an English text we choose an arbitrary letter, then the choice of the next letter is determined to a large extent by linguistic and grammatical considerations. Recent investigations of redundancies of several European languages carried out in the USSR (Lesohin, Luk'yanenkov, and Piotrovskiĭ, 1982) show the values for upper ($\bar{R}$) and lower ($\underline{R}$) bounds of redundancies in written and oral communication given in Table 4.2.

Table 4.2 Bounds on Redundancies

Type of communication	English $\bar{R}$	English $\underline{R}$	Russian $\bar{R}$	Russian $\underline{R}$	French $\bar{R}$	French $\underline{R}$	German $\bar{R}$	German $\underline{R}$	Polish $\bar{R}$	Polish $\underline{R}$
Spoken language	81.2	69.4	83.4	72.0	82.9	68.5	84.4	73.9	86.3	76.3
Literature	86.5	77.1	86.0	76.3	83.6	71.0	82.5	71.4	83.6	74.5
Business, science	92.1	82.9	90.1	83.4	90.4	83.9	88.2	79.6	89.5	83.6
Language (all)	84.5	71.9	83.6	72.1	83.4	70.6	85.1	71.4	85.0	74.7

Some tentative conclusions may be drawn from this table. The languages cited above exhibit extremely high redundancies, all approximately within the same bounds. There are, however, differences between the types of communications consistently for each of the languages studied. In particular, the language of business and science has a markedly higher redundancy, due to standardization and the repetitive nature of this type of communication. French, German, and Polish literature exhibit lower redundancy, and for the last two languages, literature is less predictable than the spoken communication. The reverse is the case for English and Russian.

Some values for corresponding entropies for French, Russian, Samoan, and English are given in Table 4.3. Note that in the Samoan language (a Polynesian language), H_1 is quite low; this is due partly to the fact that the probability distribution of letters in the Samoan language is extremely non-uniform, whereas it is relatively uniform in Russian. Another factor is the average word length: 3.2 letters in Samoan, 4.1 in English, and 5.3 in Russian. Thus, in Samoan the space character has a high probability of occurrence.

Table 4.3 Estimates of Entropies for Four Languages of Different Families (in bits/symbol)

	French	Russian	Samoan	English
H_1	3.96	4.35	3.40	4.03
H_2	3.17	3.52	2.68	3.32
H_3	2.83	3.01	2.01	3.10

Exercises

4.17   Solve the nine ball problem of Example 4.4.2 when the first weighing
        results in the tilt shown.

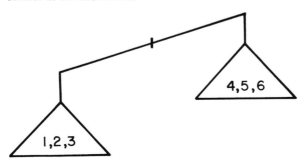

4.18   Devise an alternative procedure to solve the nine coin problem when the
        three steps yield entropies of log 3, log 2, and log 3 bits, respectively.

4.19   Show that three weighings are not in general sufficient to isolate an odd
        ball among 13 balls, even though log 26 < log 27. (Hint: Put four balls
        on each pan and observe what happens when the scales balance.)

4.20   Among 25 balls, 24 are identical and 1 is lighter. Devise a procedure to
        locate the lighter ball in three weighings. (Hint: In step 1, weigh two
        groups of 8 balls each, reserving ball 9.)

4.21   You are given a red, a white, and a blue coin, all of the same weight.
        You are also given a red, a white, and a blue coin of equal weight but
        lighter than the first three. The lighter coins appear identical to the
        heavier ones. Find the three heavier ones in two weighings. Justify
        each step in terms of the corresponding reduction in uncertainty by
        removal of information.

4.22   First, second, and third order entropies (in bits/symbol) of written
        Arabic and Portuguese are given in the following table:

	Arabic	Portuguese
H_0	4.65	4.52
H_1	4.21	3.92
H_2	3.77	3.35
H_3	2.49	3.20

        Comment on the dependencies among letters in words of written Arabic
        and Portuguese, and compare these to English.

5

DECISION MAKING IN UNCERTAINTY

To be, or not to be . . .
William Shakespeare, *Hamlet*

5.1 THE STRUCTURE OF THE PROBLEM

In a very real sense, this entire book deals with decision making in uncertainty. Whether in a casino or in a corporate board room, the rules of probability reveal information about an uncertain event, at least in the long run; such information, as we saw in Chapter 4, influences our methods of making decisions and taking risks. In this chapter we will formulate mathematical models for decision making, realizing that no single model is appropriate in all situations; rather, the choice of method may depend on characteristics of the specific problem and even on (subjective) characteristics of the decision maker's attitude toward risk. During the summer 1981 strike of the Professional Air Traffic Controllers Organization (PATCO), Robert L. Dupont (1981) reflected upon three principles that govern perception of risk.

1. Risk is perceived as greater when the decision maker has no control over the uncertain situation.
2. Risk is perceived as greater when it involves a single event (e.g., a plane crash) rather than a long run of repeated events (e.g., car crashes).
3. Risk is perceived as greater when the decision maker is unfamiliar with the situation.

While such psychological principles often affect decision making, our basis must be rational mathematical models. Our understanding of probability, then, allows us to formulate models for Dupont's "control of uncertainty" and "long run of events." Let us begin our discussion with an example.

Example 5.1.1   A drilling company owns rights in the North Sea and must decide whether to proceed to drill or to sell the drilling rights. Their cost with each decision clearly depends on the presence of oil and is assessed as follows (with figures in millions of dollars). If oil is present, drilling will result in a profit, expressed as a cost of -10, and failing to drill will result in a cost of 5, since competitors will use the oil. If no oil is present, drilling will result in a net cost of 7 and failing to drill will result in a cost of -1 (net gain), since the drilling costs can be invested elsewhere. The company has the option of performing a seismic test, at the cost of 5, which may indicate the presence of oil. Tests results can be positive, negative, or inconclusive. From past records on sites where oil eventually was found, the test gave positive results 70 percent of the time and inconclusive results 20 percent of the time; where no oil was found, the test gave 80 percent negative and 10 percent inconclusive results.

This problem setting illustrates the general structure of decision problems and the following specific elements.

Elements of Decision Making

1. A set of *states of nature* $\theta_1, \theta_2, \ldots, \theta_k$, one of which is true but unknown.
2. A set of possible actions (*decisions*) $a_1, a_2, \ldots, a_m$.
3. A *loss function* $W(\underline{\theta}, \underline{a})$ depending on the state of nature $\underline{\theta}$ and the action $\underline{a}$. In this discussion, the values of the loss function can be specified in a k $\times$ m table and we will write $W(\underline{\theta}, \underline{a})$ where $\underline{\theta} = (\theta_1, \theta_2, \ldots, \theta_k)$ and $\underline{a} = (a_1, a_2, \ldots, a_m)$.
4. An *experiment* with outcomes $z_1, \ldots, z_r$, the probabilities of which depend on the true states of nature θ_i and are denoted $f(z_j|\theta_i)$ for $i = 1, 2, \ldots, k$, and $j = 1, 2, \ldots, r$. These are conditional probabilities such that

$$\sum_{j=1}^{r} f(z_j|\theta_i) = 1 \qquad \text{for any value of i}$$

5. A *strategy* or decision rule representing a correspondence between the experimental outcomes $\{z_j\}$ and the possible actions $\{a_i\}$. Strategies are denoted by ordered r-tuples of actions $\underline{s} = (a_{i_1}, a_{i_2}, \ldots, a_{i_r})$ with a_{i_1} corresponding to outcome z_1 (i.e. $\underline{s}(z_1) = a_{i_1}$), and so on. (A decision problem with m possible actions and r experimental outcomes will have m^r different strategies.)
6. The *risk* or *average loss* associated with a strategy $\underline{s}$ and a particular state of nature θ_i given by:

$$R(\underline{s}, \theta_i) = \sum_{j=1}^{r} f(z_j|\theta_i) W(\theta_i, s(z_j)) \qquad (5.1.1)$$

for i = 1, 2, . . . , k. Explicitly, $R(\underline{s}, \theta_1)$ is the first component of the risk vector $R(\underline{s}, \underline{\theta})$, and so on. Note that $\underline{s}(z_j)$ is some action specified by strategy $\underline{s}$ when the experimental outcome is z_j.

Element 4, an experiment to gain additional sample information about the states of nature, may not be available in every decision situation. Cases without experimental data are termed no-data decisions and use nonprobabilistic criteria. If information from an experiment can be expressed in terms of probabilities, these can be used by the decision-making process. We will consider both types of situations in subsequent sections.

We illustrate the general framework with the situation of Example 5.1.1.

1. The states of nature are

    θ_1 = oil is present
    θ_2 = oil is not present $\quad$ (k = 2)

2. The possible actions are

    a_1 = drill
    a_2 = sell drilling rights $\quad$ (m = 2)

3. The loss function is as follows:

$W(\theta, \underline{a})$	a_1 drill	a_2 sell
θ_1: oil	-10	5
θ_2: no oil	7	-1

4. The seismic test (experiment) has these possible outcomes:

    z_1 = positive
    z_2 = negative $\quad$ (r = 3)
    z_3 = inconclusive

with conditional probabilities as follows:

$f(\underline{z}|\theta)$	z_1 positive	z_2 negative	z_3 inconclusive
θ_1: oil	0.7	0.1	0.2
θ_2: no oil	0.1	0.8	0.1

Note that $f(z_1|\theta_1) + f(z_2|\theta_1) + f(z_3|\theta_1) = 1$, and similarly for θ_2.

5. One possible strategy for this problem is to sell the rights if (and only if) the test results are positive. This strategy would be indicated as follows:

$$\underline{s}_1 = (a_1, a_2, a_2)$$

or

$$\underline{s}_1(z_1) = a_1, \qquad \underline{s}_1(z_2) - a_2, \qquad \text{and} \qquad \underline{s}_1(z_3) = a_2$$

Another possible strategy is $\underline{s}_2 = (a_1, a_1, a_1)$, a constant strategy in which we always drill regardless of the test results. The other possible strategies (out of the possible $m^r = 2^3 = 8$ in this case) are

$$\underline{s}_3 = (a_1, a_1, a_2)$$

$$\underline{s}_4 = (a_1, a_2, a_1)$$

$$\underline{s}_5 = (a_2, a_1, a_1)$$

$$\underline{s}_6 = (a_2, a_1, a_2)$$

$$\underline{s}_7 = (a_2, a_2, a_1)$$

$$\underline{s}_8 = (a_2, a_2, a_2)$$

6. The risk components associated with $\underline{s}_1$ are

$$R(\underline{s}_1, \theta_1) = (0.7)(-10) + (0.1)(5) + (0.2)(5) = -5.5$$

$$R(\underline{s}_1, \theta_2) = (0.1)(7) + (0.8)(-1) + (0.1)(-1) = -0.2$$

The risks of the complete set of strategies are given in Table 5.1.

Notice that both components of $R(\underline{s}_5, \theta)$ and $R(\underline{s}_7, \theta)$ are at least as large as the corresponding components of $R(\underline{s}_1, \theta)$. Thus, it seems unreasonable ever to use $\underline{s}_5$ or $\underline{s}_7$, since $\underline{s}_1$ outperforms them, on the average, no matter what the value of θ is. We call $\underline{s}_5$ and $\underline{s}_7$ *inadmissible strategies* and say that $\underline{s}_1$ *dominates* $\underline{s}_5$ and $\underline{s}_7$ (or $\underline{s}_5$ and $\underline{s}_7$ are dominated by $\underline{s}_1$). Similarly, $\underline{s}_3$ is inadmissible, since it is dominated by $\underline{s}_4$, and $\underline{s}_8$ is inadmissible, since it is dominated by $\underline{s}_6$. We have this general definition:

> A strategy $\underline{s}$ is *inadmissible* if there exists a strategy $\underline{t}$ such that
>
> $$L(\underline{s}, \theta_i) \geqslant L(\underline{t}, \theta_i)$$
>
> for all $i = 1, \ldots, k$, with strict inequality for at least one i.

In a given class of strategies one might have several admissible ones, none of which is the best, in the sense that some are better for one state of nature and the other may yield a smaller risk for a second state of nature, as Table 5.1 illustrates. Consider a man (see L. J. Savage, 1972) who chose chicken

Table 5.1   Risks of Strategies Associated with Example 5.1.1

$R(\underline{s}, \theta)$	$\underline{s}_1$	$\underline{s}_2$	$\underline{s}_3$	$\underline{s}_4$	$\underline{s}_5$	$\underline{s}_6$	$\underline{s}_7$	$\underline{s}_8$
θ_1	-5.5	-10	-7	-8.5	0.5	3.5	2	5
θ_2	-0.2	7	6.2	0.6	6.2	-5.8	-0.2	-1

when offered chicken or steak. He was then told that lobster was also on the menu, at which point he replied, "In that case, I'll have steak." A rational explanation of his behavior illustrates the concept of dominant strategies. If in this situation steak is indistinguishably better than chicken and chicken is indistinguishably better than lobster, but steak is distinguishably better than lobster, we have the following diagram:

Here steak is logically better than chicken, but perhaps not distinguishably so. When the man considered only chicken and steak, he could see no difference and chose chicken, virtually at random; the merit of lobster lies just far enough away to resolve this quandary.

Exercises

5.1   For a three-state two-action decision problem, the losses and probability distributions of the outcomes of the experiment are as follows:

Losses

$W(\theta, \underline{a})$	a_1	a_2
θ_1	4	1
θ_2	3	5
θ_3	2	6

Probabilities

$f(\underline{z}|\theta)$	z_1	z_2
θ_1	0.7	0.3
θ_2	0.5	0.5
θ_3	0.4	0.6

List all possible strategies and their risk components. Identify any inadmissible strategies.

5.2   We are faced with the decision of investing in the development of either a coal plant or a nuclear plant. Our profits will depend on the discount rate, which will be either 20 percent or 18 percent. Net revenues are given as follows:

Revenue		Discount rate	
		20%	18%
Industry	Coal	14	15
	Nuclear	10	20

For further information, we ask a group of financial experts to predict trends in discount rates. In past cases, the experts' response record has been:

		Experts' responses		
		20%	18%	Inconclusive
Actual rate	20%	0.7	0.1	0.2
	18%	0.1	0.7	0.2

(For example, in 70% of the cases, the experts predicted correctly the 20% discount rate.) Identify the six elements of this decision problem. Find the risks of all strategies. Which strategies are admissible?

5.3 A recent study with questionable statistical basis has claimed to link coffee consumption and incidence of cancer (see MacMahon et al., 1981). You have just poured your morning coffee when it occurs to you to weigh the possible consequences; your relative preferences are represented in the following loss table.

		Actions	
		a_1 drink coffee	a_2 pour coffee out
Effect of coffee	θ_1: harmless	0	1
	θ_2: harmful	10	1

You ask the opinion of a medical expert who has made several studies on cancer incidence in the past with the following conditional probabilities of being correct:

		Expert concludes	
		Harmful	Harmless
Coffee is	harmless (θ_1)	0.1	0.9
	harmful (θ_2)	0.8	0.2

(That is, in 10% of the cases when an agent is actually harmless, the expert has concluded a relation to cancer.) Compute the risks of the strategies in this decision and identify the admissible ones.

5.4 In analyzing childhood behavior, A. Wuffle (1979) has presented the following model. Suppose that a child and his or her parents are in disagreement about something that should be done. Take the example in which Suzy's father tells her to go practice her violin and Suzy would prefer to play outside instead. Let us assume that Suzy is more displeased at practicing than she is pleased at playing outside and that her choice of obeying or disobeying her father is somewhat dependent on whether he monitors what she does. This model of behavior may be represented in the following loss table (note that a loss of -1 represents a gain of $+1$, that is, a reward).

		Suzy's action	
		Obey (a_1)	Disobey (a_2)
Father	monitors (θ_1)	-1	6
	does not monitor (θ_2)	5	-2

Suzy replies to her father, "I will in a minute," to try to determine if he is going to monitor her behavior (this is the experiment). From past experience, she knows that his answer will be one of:

z_1: silence
z_2: "OK"
z_3: "Suzy, do what I told you."

The conditional probabilities of θ_1 and θ_2 in each of the cases are given as follows:

	Father's answer		
	z_1	z_2	z_3
Father monitors (θ_1)	0	0.4	0.6
Does not monitor (θ_2)	0.5	0.5	0

a. List Suzy's possible strategies in deciding whether to practice her violin.
b. How would this model of behavior be changed in the absence of rewards or punishments from Suzy's father?

5.2 MINIMAX STRATEGIES

Let us take a pessimistic (or conservative) view (see Table 5.2) of the risks of the drilling problem given in Table 5.1 and mark for each strategy the largest risk components. Now, the maximum risk is smallest for strategy s_1, so we call s_1 the minimax strategy:

> A *minimax* strategy s among all possible strategies s_j is one for which $\max_{1 \leq i \leq k} R(s, \theta_i)$ is minimized.

Table 5.2 Maximum Risks of Drilling Example

$R(s, \theta)$	s_1	s_2	s_3	s_4	s_5	s_6	s_7	s_8
θ_1	-5.5	-10	-7	-8.5	0.5	3.5	2	5
θ_2	-0.2	7	6.2	0.6	6.2	-5.8	-0.2	-1
$\max R(s, \theta_i)$	-0.2	7	6.2	0.6	6.2	3.5	2	5

A minimax strategy is extremely cautious, since it hedges against the worst that might happen. It reminds us of the king who starved to death because he feared his food was poisoned and would not trust even his official tasters. A businessperson who used minimax strategies would soon be out of business, since almost any venture might lose money whereas sitting tight incurs no losses—and no profits. However, minimax is useful in certain situations. Consider Example 5.2.1, in which no experiment is carried out and we have only the loss table (a no-data decision problem).

Example 5.2.1 You and I play a game in which we each have three sticks: one yellow, one red, and one blue. Each of us chooses a stick, not seeing the other's choice. When the choices are revealed, you receive or give up an amount of money according to Table 5.3. Here, the states of nature θ_i are my selections, the actions a_j are your selections, and Table 5.3 gives values of $W(\underline{\theta}, \underline{a})$. Since no experiment is carried out, this is a no-data decision problem, and the minimax criterion is applied directly to the loss table. Your minimax action is the action a_t ($t = 1, 2, 3$) for which $\max_{\underline{\theta}} W(\underline{\theta}, a_t)$ is a minimum:

a_t	$\max_{\underline{\theta}} W(\underline{\theta}, a_t)$
a_1	15
a_2	-2
a_3	10

Hence, red is your minimax action, by which you assure yourself at least $2. Looking at the game from my perspective (with your choices assuming the role of states of nature), Table 5.3 becomes a table of *utilities* rather than losses. My minimax action (in this case called *maximin*) would maximize the smallest values of a column; if I choose yellow, I can be sure that you will get no more than $2. If we both choose our actions using the minimax criterion, then I will end up paying you $2. In fact, if you use minimax and I do not, then I will be penalized. Here, -2 is called a *saddle point*, or *equilibrium point*, of the loss matrix; it is simultaneously the minimum of a row and the maximum of a column.

Table 5.3 Your Losses in a Game (Negative Amounts Indicate a Win)

	$W(\underline{\theta}, \underline{a})$	My actions		
		Yellow (θ_1)	Red (θ_2)	Blue (θ_3)
	Yellow (a_1)	3	-30	15
Your actions	Red (a_2)	-2	-20	-3
	Blue (a_3)	4	10	-20

Table 5.4 Opportunity Loss for Game of Example 5.2.1

Opportunity loss	θ_1	θ_2	θ_3
a_1	5	0	35
a_2	0	10	17
a_3	6	40	0

If you as a decision maker are concerned about what might have been, then your criterion might be to minimize the maximum loss of opportunity. A table of *opportunity loss* is obtained from the loss table by subtracting each entry from the smallest entry in its state of nature. This represents the difference in value between what one obtains for a given action and state and what one would obtain if one knew beforehand that the given state was the true state. For the game of Table 5.3, opportunity loss from your perspective is given by Table 5.4. For example, if you take action a_3 and θ_1 is the true state of nature, then your opportunity loss is $4 - (-2) = 6$, since you could have been $6 better off had you known beforehand that θ_1 was the true state. Now, let's look at the maximum opportunity loss (O. L.) for each action:

	$\max_{\underline{\theta}}$ O. L.	
a_1	35	
a_2	17	← minimum
a_3	40	

We would select action a_2, since its greatest loss is smaller than the greatest loss of either a_1 or a_3. Selecting an action according to the criterion of opportunity loss is analogous to the minimax procedure of Example 5.2.1, except that the loss table $W(\theta, \underline{a})$ is replaced by a table of *conditional* losses.

Any minimax-type procedure (applied to loss or opportunity loss) is very conservative (cautious), but perhaps the most serious criticism is its sensitivity to irrelevant alternatives; consider the following example.

Example 5.2.2 You select one of two actions a_1 or a_2 and then a coin is flipped; you receive a payoff according to the following table:

Payoff		Your action	
		a_1	a_2
Coin lands:	heads	100	40
	tails	0	50

Since these values are payoffs, the maximin criterion would pick the action that maximizes the minimum value of each column; here, a_2. The opportunity loss table is obtained by subtracting each value from the largest value in its row:

Opportunity loss		Your action	
		a_1	a_2
Coin lands	heads	0	60
	tails	50	0

Minimax loss selects action a_1, since its maximum loss (\$50) is less than the maximum loss of a_2 (\$60). Now suppose we add a new action a_3 with payoffs of $-\$500$ (a loss to you) and \$65:

Adjusted payoffs	a_1	a_2	a_3
Heads	100	40	-500
Tails	0	50	65

Opportunity loss becomes:

Adjusted opportunity loss	a_1	a_2	a_3
Heads	0	60	600
Tails	65	15	0

Now minimax loss selects action a_2 since 60 is less than 65 and 600. Although a_3 was not selected, its mere presence altered the decision from a_1 to a_2.

Exercises

5.5 In Example 5.2.1, the pair of actions (you pick red, I pick yellow) is an equilibrium or saddle point. Verify that no other pair is a saddle point.

5.6. a. In the following loss table, what is the minimax action? Would you recommend its use?

$W(\underline{\theta}, \underline{a})$	a_1	a_2
θ_1	15	1
θ_2	25	2
θ_3	30	30.1

b. Construct the opportunity loss table for this decision. What is the minimax decision in this case?

5.7 What is the minimax strategy for the coal-nuclear problem of Exercise 5.2?

5.8 In the coffee decision of Exercise 5.3, what decision does minimax loss select? Interpret the decision in this context.

5.9 a. For the childhood behavior model of Exercise 5.4a, what action does minimax loss select? What is the practical interpretation of this criterion for this situation?

b. What action would Suzy take using the minimax criterion as applied to opportunity loss?

5.10 You plan to buy a $2 win ticket for a three-horse race. According to the quoted odds, your potential profit is

Profit	Your bet		
	1	2	3
1	4	-2	-2
Winner 2	-2	1.5	-2
3	-2	-2	0.7

What horse does minimax loss select?

5.3 BAYESIAN DECISION MAKING

In addition to the sample information available from an experiment (element 4 of Section 5.1), we might have other subjective information about the decision problem that reasonably should be incorporated into our decision-making process. In the classical sense (recall Section 1.2), information cannot be regarded as probabilities unless it results from random sampling carried out on a well-defined sample space. However, the subjective interpretation of probability (Section 1.5) allows us to judge a situation by a degree of belief which may be regarded as a probability. Then Bayes' rule (Section 2.2) allows us to revise our subjective probabilities based on the sample information we subsequently obtain. Example 5.3.1 illustrates how these ideas fit into our decision-making framework.

Example 5.3.1 In the problem setting of Example 5.1.1, a company must decide whether to drill for oil. Based on past drilling experiences but no scientific sampling, the company feels that the chances of finding oil in this location are about 60 percent; this represents a degree of belief, a subjective probability, Now, the information from the seismic test may cause the company to alter these prior probabilities $P(\text{oil}) = P(\theta_1) = 0.6$ and $P(\text{no oil}) = P(\theta_2) = 0.4$. The average loss with respect to these prior probabilities is called Bayesian loss.

The *Bayesian loss* for an action a_ϱ in a no-data decision problem with prior probabilities $p(\theta_i)(i = 1, \ldots, k)$ and loss $W(\theta_i, a_\varrho)$ is

$$B(a_\varrho) = \sum_{i=1}^{k} p(\theta_i) W(\theta_i, a_\varrho) \qquad (5.3.1)$$

We write $B(\underline{a}) = [B(a_1), \ldots, B(a_m)]$.

Equation (5.3.1) for $B(a_\varrho)$ is reminiscent of Equation (5.1.1) for the risk $R(\underline{s}, \underline{\theta})$. Note, however, that R is a risk vector with k components, each of which is an average of the loss W with respect to probabilities of experimental outcomes z_j. $B(\underline{a})$ is a vector of m components, each of which is an average of the loss W with respect to prior probabilities of the states of nature θ_i. In this example, we compute

Action, a	Bayesian loss, B(a)
a_1 : drill	$(0.6)(-10) + (0.4)(7) = -3.2$
a_2 : sell	$(0.6)(5) + (0.4)(-1) = 2.6$

The (prior) Bayesian decision would be to drill (action a_1) because we expect smaller loss with this choice. If, in addition, we have experimental data, Bayes' rule allows us to compute posterior probabilities based on the additional information, in the manner of Section 2.2.

State of nature θ_i	Prior probability $P(\theta_i)$	Sample information $f(z_j \mid \theta_i)$	Joint probability $f(z_j \cap \theta_i)$	Posterior probability $f(\theta_i \mid z_j)$	
θ_1 : oil	0.6	0.7	0.42	0.91	
θ_2 : no oil	0.4	0.1	0.04	0.09	z_1
			0.46	1.00	
θ_1	0.6	0.1	0.06	0.16	
θ_2	0.4	0.8	0.32	0.84	z_2
			0.38	1.00	
θ_1	0.6	0.2	0.12	0.75	
θ_2	0.4	0.1	0.04	0.25	z_3
			0.16	1.00	

Having first computed the posterior probabilities $f(\theta_i \mid z_j)$, the expected loss associated with each action is computed with respect to these revised probabilities.

In a decision problem with data the *Bayesian loss* for an action a_ϱ and an experimental outcome z_j is the expected value of the loss function $W(\theta, a_\varrho)$ with respect to posterior probabilities $f(\theta | z_j)$. We write

$$B(a_\varrho | z_j) = \sum_{i=1}^{k} f(\theta_i | z_j) W(\theta_i, a_\varrho) \qquad (5.3.2)$$

for $j = 1, 2, \ldots, r$ and $\varrho = 1, 2, \ldots, m$.

Note the similarity of Equation (5.3.2) to its analog. Equation (5.3.1) for the no-data problem; we merely replace $p(\theta_i)$ with our revised probability $f(\theta_i | z_j)$. In the case of Example 5.3.1, computations yield:

$B(\underline{a}, \underline{z})$	z_1	z_2	z_3
a_1	$(0.91)(-10)$ $+ (0.09)(7)$ $= -8.47$	$(0.16)(-10)$ $+ (0.84)(7)$ $= 4.28$	$(0.75)(-10)$ $+ (0.25)(7)$ $= -5.75$
a_2	$(0.91)(5)$ $+ (0.09)(-1)$ $= 4.46$	$(0.16)(5)$ $+ (0.84)(-1)$ $= -0.04$	$(0.75)(5)$ $+ (0.25)(-1)$ $= 3.5$

Thus the Bayesian strategy is $\underline{s} = (a_1, a_2, a_1)$; that is, $\underline{s}(z_1) = a_1 = \underline{s}(z_3)$ and $\underline{s}(z_2) = a_2$, since these actions yield smaller Bayesian losses for each outcome z_j. To compute the risk of such a Bayesian strategy, we analyze each component separately, not using the Bayesian losses.

The component of risk for a Bayesian strategy $\underline{s} = (a_{\varrho_1}, a_{\varrho_2}, \ldots, a_{\varrho_r})$ and a state of nature θ_i is given by

$$R(\underline{s}, \theta_i) = \sum_{j=1}^{r} f(z_j | \theta_i) W(\theta_i, \underline{s}(z_j)) \qquad (5.3.3)$$

The *Bayesian risk* is then

$$R(\underline{s}) = \sum_{i=1}^{k} p(\theta_i) R(\underline{s}, \theta_i) \qquad (5.3.4)$$

Equation (5.3.3) is the ordinary risk of Equation (5.1.1) where the strategy $\underline{s}$ is a Bayesian one. The Bayesian risk of Equation (5.3.4) is analogous to the Bayesian loss of Equation (5.3.1) with the loss function $W(\theta_i, a_\varrho)$ replaced by the risk component $R(\underline{s}, \theta_i)$.

For example 5.3.1, the risk components of the Bayesian strategy $\underline{s} = (a_1,$ $a_2, a_1)$ are

$$R(\underline{s}, \theta_1) = (-10)(0.7) + (5)(0.1) + (-10)(0.2) = -8.5$$

$$R(\underline{s}, \theta_2) = (7)(0.1) + (-1)(0.8) + (7)(0.1) = 0.6$$

and the Bayesian risk is

$$R(\underline{s}) = (-8.5)(0.6) + (0.6)(0.4) = -4.86,$$

which is an average of the actual values from the loss table on p. 111.

We have led you through this analysis without discussing how the prior probabilities $p(\underline{\theta})$ are chosen. In practice, this is a difficult question, although an important one, since the optimal Bayes decision is quite sensitive to our choice of priors. In Example 5.3.1, the Bayesian decision (with priors 0.6 and 0.4) for the no-data decision problem was to drill for oil. In Exercise 5.11, you are asked to show that priors of $p(\theta_1) = 0.3$ and $p(\theta_2) = 0.7$ reverse the Bayesian decision and lead us to sell the drilling rights!

Exercises

5.11   Suppose that the prior probabilities in Example 5.3.1 are $P(\text{oil}) = 0.3$ and $P(\text{no oil}) = 0.7$. Show that the Bayesian decision for the no-data problem is to sell the drilling rights. What value for $P(\text{oil})$ would cause us to be indifferent to whether we drill or sell the drilling rights?

5.12   Mary Doe smokes and is trying to decide whether to stop. Her assessments of the consequences of various decisions are given in the following loss table:

		Mary's actions	
		Stop smoking (a_1)	Continue smoking (a_2)
	$W(\underline{\theta}, \underline{a})$		
States of nature	Smoking reduces life expectancy (θ_1)	0	1
	Smoking doesn't reduce life expectancy (θ_2)	0.8	0.1

Mary's prior probability that smoking is harmful is 0.4. If she is a Bayesian decision maker, should she stop smoking? For what range of prior probabilities, $p(\theta_1)$, would she continue to smoke?

5.13   What is the Bayesian decision for Exercise 5.1 if the states of nature are equally likely?

5.14   Show that a Bayesian strategy with strictly positive priors and equal risk components is a minimax strategy. (Hint: Assume that this is not true and exhibit a contradiction.)

5.15 In Exercise 5.2, what Bayesian action will you take if the experts predict a 20 percent discount rate and the prior probabilities on the states of nature are equal?

5.16 Compute the Bayesian strategy for Exercise 5.3, if you believe coffee is much more likely to be harmless than harmful at odds of 4:1. What is the Bayesian risk of this strategy?

5.17 In Exercise 5.4a, suppose that Suzy's father is more likely to monitor her action with $p(\theta_1) = 0.7$. Compute her Bayesian strategy and the associated Bayesian risk.

5.18 A local politician is being urged by the press to make a statement about whether she will run for national office. In an upcoming news conference, she may take one of three courses of action:

a_1 : announce that she will not run
a_2 : refuse to comment on her candidacy
a_3 : announce that she will run

The effect of each possible announcement is dependent on the value of π, the proportion of votes she will receive in an upcoming election. Her advisors define three ranges of the unknown value π:

θ_1: $\pi < 40\%$
θ_2: $40\% \leqslant \pi < 60\%$
θ_3: $\pi \geqslant 60\%$

The consequences of each announcement for a given state of nature are given in the following loss table.

$W(\underline{\theta}, \underline{a})$	a_1	a_2	a_3
θ_1	0	2	4
θ_2	5	3	5
θ_3	10	9	6

The politician decides to conduct an opinion poll, thinking that the value of the sample proportion in her favor, p, might give some information about the unknown value π. From statistical analysts she find that polls of this sort in the past have given information with the following relative frequencies:

	z_1	z_2	z_3	z_4
$f(z|\theta)$	$0 \leqslant p < 0.25$	$0.25 \leqslant p < 0.50$	$0.50 \leqslant p < 0.75$	$0.75 \leqslant p \leqslant 1.00$
θ_1	0.5	0.4	0.1	0
θ_2	0.2	0.5	0.2	0.1
θ_3	0	0.2	0.5	0.3

a. List five strategies and evaluate their average losses.
b. Which is the best of these strategies if the politician has priors $P(\theta_1) = 0.3$ and $P(\theta_2) = 0.5$?
c. What should she say at the news conference if 18 out of 40 voters polled express a preference for her?

5.19   A wildlife preserve has a small number A of deer. You know that A is 1, 2, or 3, and you assign probabilities of 0.2, 0.2, and 0.6, respectively, to these values. You catch a deer at random from the preserve, tag it, and let it go. The next day you again catch a deer, note whether it is tagged, and let it go. You must then decide whether the preserve has three deer, winning a prize if your decision is correct. What is the best method of deciding? (Hint: Construct a "one-zero" utility table for two states of nature: less than 3 deer (θ_1) and 3 deer (θ_2).)

5.20   The server, against his perennial tennis opponent, has two serves in his arsenal; a hard one that is very effective if it lands and a soft one that lands with greater reliability but is not as effective when it does. Assume that the hard serve lands fairly with probability 1/2 and that when it does the server has a 3/4 chance of winning the point. Assume that the soft serve lands fairly 3/4 of the time and let p be the probability that if the soft serve lands safely the server will win the point. With the usual tennis allowance of two serves, the server has four possible serving strategies, denoted by HH, HS, SH, and SS. For what ranges of p are each of the four serving strategies optimal? [Hint: For each of the four strategies, the probability of success will be the sum of the probabilities of (1) winning on the first serve and (2) losing on the first serve and winning on the second.]

5.4 RANDOMIZED DECISION RULES

The process by which we make decisions is a complicated one and does not lend itself readily to mathematical modeling. For example, the drilling company of Example 5.1.1 might decide to drill if the seismic test yields positive results and sell if results are negative, but it might be quite unsure about what action to take if the results are inconclusive. The decision maker will vacillate from one choice to the other and finally make a decision, even then not convinced that it was the best one. One attempt to model this complicated process for analysis is through *randomized decision rules*, also called *mixed strategies.*

Example 5.4.1   Let us look again at the no-data decision for the drilling company with loss specified by the loss table on p. 111. Let us suppose that the decision maker rolls a fair die: If an ace comes up, the company will drill: otherwise it will sell the drilling rights. Thus we have a randomized action a*, with

$$a^* = \begin{cases} a_1 & \text{with probability } 1/6 \\ a_2 & \text{with probability } 5/6 \end{cases}$$

In a no-data decision problem with actions $a_1, \ldots, a_m$ we can form a *randomized action* a* by taking a* = a_j with probability p_j where

$0 \leqslant p_i \leqslant 1$ and $p_1 + p_2 + \cdots + p_m = 1$. The risk when θ_j is the true state of nature is

$$R(a^*, \theta_j) = \sum_{i=1}^{m} (p_i) W(\theta_j, a_i)$$

Note that a^* is thus a random variable taking values $a_1, \ldots, a_m$ with probabilities $p_1, \ldots, p_m$, respectively.

In our drilling context $p_1 = 1/6$ and $p_2 = 5/6$ ($m = 2$), and our expected loss components are

$$R(a^*, \theta_1) = (1/6)(-10) + (5/6)(5) = 2.5$$

$$R(a^*, \theta_2) = (1/6)(7) + (5/6)(-1) = 0.33$$

the maximum expected loss with a^* being 2.5. For a second randomized action

$$b^* = \begin{cases} a_1 & \text{with probability } = 1/4 \\ a_2 & \text{with probability } = 3/4 \end{cases}$$

the risk components are $R(b^*, \theta_1) = 1.25$ and $R(b^*, \theta_2) = 1$, the largest being 1.25. Thus we see that a^* cannot be the best randomized action in a minimax sense.

In a two-action ($m = 2$) decision problem, graphs are helpful in determining optimal randomized actions. Let p denote the probability that we select a_1, say; then, clearly, we use a_2 with probability $1 - p$ and $R(a^*, \underline{\theta})$ is a function of p. Here,

$$R(a^*, \theta_1) = (p)(-10) + (1 - p)(5) = 5 - 15p$$

$$R(a^*, \theta_2) = (p)(7) + (1 - p)(-1) = 8p - 1$$

Figure 5.1 graphs these two linear functions of p. The boldface line is the maximum expected loss and it attains its minimum at the point p where the curves intersect. Thus the value of p giving a minimax randomized decision is found as follows:

$$5 - 15p = 8p - 1$$

i.e. $6 = 23p$

or $p = \dfrac{6}{23}$

Namely, we choose action a_1 with probability 6/23 and a_2 with probability 17/23. The expected loss with this value is

$$5 - (15)(6/23) = (8)(6/23) - 1 = 25/23 = 1.087$$

the same value for each state of nature. Compare this value with the individual entries of the loss table on p. 111.

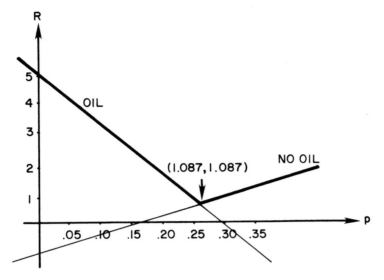

Figure 5.1   Risks of randomized actions of Example 5.4.1.

Now let us consider the addition of sample information (an experiment) to our randomized decision problem.

Example 5.4.2   The decision maker in the drilling company uses the results of the seismic test (the probabilities given in the conditional probability table on p. 111) as follows: if test is positive, drill; if test is negative, sell the rights; if the test is inconclusive, toss a fair coin and drill if heads, sell if tails. If we call this randomized strategy $\underline{s}^*$, we can relate it to the pure (nonrandomized) strategies of the loss table on p. 111. If the toss is heads, then $\underline{s}^* = (a_1, a_1, a_2) = \underline{s}_3$; if tails, $\underline{s}^* = (a_1, a_2, a_2) = \underline{s}_1$. Thus

$$\underline{s}^* = \begin{cases} \underline{s}_1 & \text{with probability } 1/2 \\ \underline{s}_3 & \text{with probability } 1/2 \end{cases}$$

In a decision problem with pure strategies $\underline{s}_1, \ldots, \underline{s}_{m^r}$ a randomized strategy $\underline{s}^*$ is formed by taking

$$\underline{s}^* = \underline{s}_\varrho \qquad \text{with probability } p_\varrho$$

where $0 \leqslant p_\varrho \leqslant 1$ and $p_1 + \cdots + p_{m^r} = 1$. The expected risk with $\underline{s}^*$ is

$$R(\underline{s}^*, \theta_i) = \sum_{\varrho=1}^{m^r} (p_\varrho) R(\underline{s}_\varrho, \theta_i)$$

$$= \sum_{\varrho=1}^{m^r} (p_\varrho) \left[\sum_{j=1}^{r} f(z_j | \theta_i) W(\theta_i, \underline{s}_\varrho(z_j)) \right]$$

In Example 5.3.2, $p_1 = 1/2$, $p_3 = 1/2$, all other p_i's are 0, so the risk components of $\underline{s}^*$ (using the conditional probability table on p. 111) are

$$R(\underline{s}^*, \theta_1) = (1/2)(-5.5) + (1/2)(-0.2) = -2.85$$

$$R(\underline{s}^*, \theta_s) = (1/2)(-7) + (1/2)(6.2) = -0.4$$

Now suppose the decision maker considers some method of randomization other than a coin toss, in the case of inconclusive results from the seismic test. Let p be the probability with which the randomization indicates selling; we have a set of possible mixed strategies $\{\underline{s}^*\}$ with risk components

$$R(\underline{s}^*, \theta_1) = pR(\underline{s}_1, \theta_1) + (1 - p)R(\underline{s}_3, \theta_1)$$

$$= p(-5.5) + (1 - p)(-7) = 1.5p - 7$$

$$R(\underline{s}^*, \theta_2) = pR(\underline{s}_1, \theta_2) + (1 - p)R(\underline{s}_3, \theta_2)$$

$$= p(-0.2) + (1 - p)(6.2) = 6.2 - 6.4p$$

The minimax randomized strategy may now be found as in Figure 5.1 at the intersection of the lines $1.5p - 7$ and $6.2 - 6.4p$.

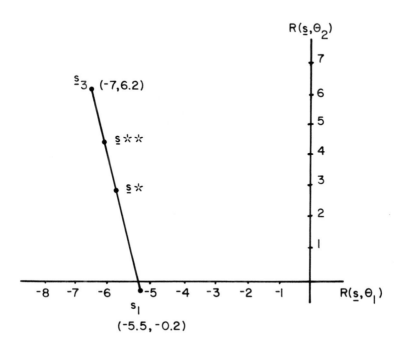

Figure 5.2   Risks of two pure strategies of Example 5.4.2.

Now, let us consider the choice of strategies over which to randomize. Graphically, we can represent a pure strategy $\underline{s}$ as a point with coordinates $(R(\underline{s}, \theta_1), R(\underline{s}, \theta_2))$. Figure 5.2 illustrates this for $\underline{s}_1$ and $\underline{s}_3$. A mixed strategy $\underline{s}^*$ is a point on the line segment joining the points $\underline{s}_1$ and $\underline{s}_3$. With $\underline{s}^*$ at the midpoint of this line segment, equal weights will be given to $\underline{s}_1$ and $\underline{s}_3$. A strategy $\underline{s}^{**}$ with $p = 1/4$ will have risk midway between the risks of $\underline{s}^*$ and $\underline{s}_3$, closer to $\underline{s}_3$ than $\underline{s}_1$ since $\underline{s}_3$ receives the greater weight $(1 - p) = 3/4$. In general, between any two pure strategies we have a dense set of mixed strategies whose risk components are on the line segment joining the pure risk components.

Let us now add a third strategy $\underline{s}_5$ to our consideration. Figure 5.3 shows the risk components of $\underline{s}_1, \underline{s}_3$, and $\underline{s}_5$. The resulting triangle, its interior together with its boundary, forms a *convex set*, that is, a set of points which will contain the entire line segment joining any two of its points. Figure 5.4 shows all of the points of Table 5.1 and reveals some interesting characteristics of mixed strategies. The pure strategies $\underline{s}_1, \underline{s}_3$, and $\underline{s}_7$ are inside the convex set of strategies and can thus be represented as a mixture of other pure strategies. Admissible strategies are the vertices of this convex set. Note that a strategy that is

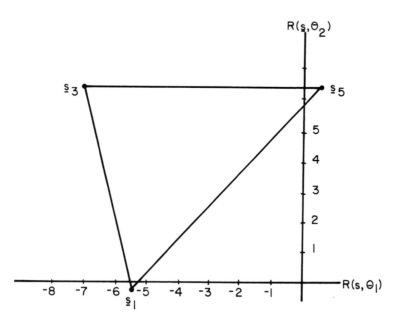

Figure 5.3   Risks of three pure strategies of Example 5.4.2.

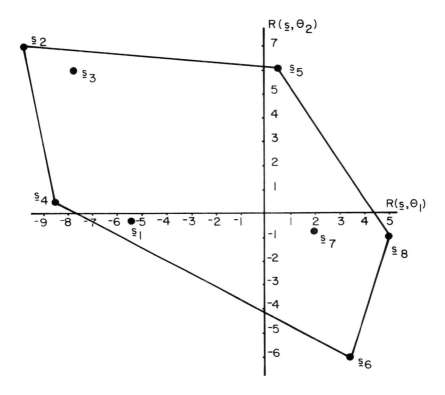

Figure 5.4   Risks of all pure strategies of Example 5.4.2.

admissible when we consider only pure strategies may become inadmissible among all randomized strategies (in Figure 5.5, strategy $\underline{s}^*$ clearly dominates strategy $\underline{s}_2$).

Hence this randomization process potentially produces strategies more appealing than pure strategies (in terms of risk) by smoothing the extreme values of the risk components of competing states of nature. However, as in any situation where the criterion is optimizing an expectation, the value of randomization is realized only in repeated applications of the decision-making process under similar circumstances.

Randomized decision procedures can be extended to the Bayesian case incorporating prior probabilities. The computations for this case are straightforward but tedious. The interested reader is referred to Weiss (1961). For further discussion of concepts developed in the last sections we refer the reader to the pioneering work of Chernoff and Moses (1959).

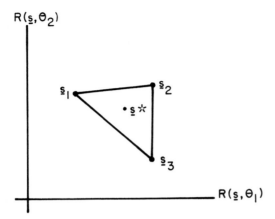

Figure 5.5   Risks of pure and randomized strategies of Example 5.4.2.

Thus far in this chapter, we have investigated a framework through which several aspects of decision making can be modeled. When people make risky choices, they sometimes act other than objectively. Experimental surveys have indicated that such departures from objective decision making tend to follow regular patterns that can be described mathematically. Kahneman and Tversky (1982) consider a variety of examples of discrepancies between subjective and objective conceptions of decisions. The concept of psychology of preferences originated in 1738 with the mathematician Daniel Bernoulli (1700–1782)—nephew of James Bernoulli (1654–1705), a founder of modern probability theory—who discussed the risk-averse characteristics of human behavior. For example, given a choice between a certain gain of $80 and a risky option offering an 85 percent chance of winning $100 and a 15 percent chance of winning nothing, most people prefer the certain gain to the gamble, despite its lower expectation. Many economic theories assume that a consumer prefers a risky venture only when its expected profit is high enough to compensate for the risk. However, psychology indicates that risk-seeking preferences are common in the choice between a sure loss and a risky chance of losing nothing.

Furthermore, survey evidence indicates that decision weights which multiply the values of outcomes do not always coincide with probabilities. If you could improve your chances to win a desirable prize, would you pay as much to raise your chance of winning from 30 percent to 40 percent as you would to raise your chance from 90 percent to certainty? Generally, an increase from 30 percent to 40 percent is perceived as less valuable than an increase from 90 percent to certainty or than an increase from impossibility to 10 percent! This inflated view of small probabilities contributes to the appeal of lottery tickets and insurance policies.

Decision making in consumer behavior is often dominated by a framing effect. Suppose you are about to buy a personal computer for $2000 and a space-game program for $100, when the salesperson tells you the program is on sale for $90 at another branch which is 30 minutes away. People are more likely to make the trip to the other store in this case than if the program were $100 and the computer were $2000 at the first store and $1990 at the branch. The total purchase and the consequences (driving 30 minutes to save $10) are the same in both versions, but the same reduction is more impressive relative to a smaller price. Thus we may buy new furniture more readily at the same time as we buy a new house, and we may purchase a car with a tape deck when it seems extravagant to add one to our present car.

The influence of a reference point on a decision may depend on an act of imagination by which one creates alternative realities. For example, imagine that the winning number in a lottery is 40356. Peter holds 56831; Emily holds 50356; and Michael holds 40357. How would you rank the ticket holders in terms of their frustration?

Exercises

5.21   Relate the computations of the oil drilling Example 5.3.1 to a Bayesian decision with general prior probabilities p and $1 - p$ (for $0 < p < 1$).

5.22   For the coal-nuclear decision of Exercise 5.2, draw the convex set of all strategies and determine the admissible ones.

5.23   Consider the following loss table and experimental probabilities [based on a discussion by Chernoff and Moses (1959)].

$W(\theta, a)$	a_1	a_2	a_3	$f(z\vert\theta)$	z_1	z_2	z_3
θ_1	0	1	3	θ_1	0.6	0.25	0.15
θ_2	3	1	0	θ_2	0.2	0.3	0.5

Determine admissible strategies from the convex set of strategies.

5.24   For the decision problem of Exercise 5.23, compute the risk components of the following strategies $\underline{s}$ and $\underline{t}$ and identify them on the convex set of strategies.

    $\underline{s}$: If z_1, then use a_1 with probability (w.p.) $1/3$ and a_2 w.p. $2/3$
       If z_2, then use a_2 w.p. $2/5$ and a_3 w.p. $3/5$
       If z_3, then use a_3 w.p. 1

    $\underline{t}$: If z_1, then use a_1 w.p. $5/6$ and a_2 w.p. $1/6$
       If z_2, then use a_1 w.p. 1
       If z_3, then use a_1 w.p. $1/10$ and a_2 w.p. $9/10$

5.25   For the coffee decision of Exercise 5.3, propose a suitable randomized strategy and evaluate its risk. Would you advocate the use of mixed strategies in this case?

5.26   On a recent trip to England, you are informed that when it rains in
London, it really rains. In fact, it seems that on rainy days it begins
pouring at 9 a.m. and continues until midnight. Therefore, before
dressing for a day in the city, you refer to a rain indicator mounted
outside your hotel room, which can indicate fair weather (z_1), wet
weather (z_2), or inconclusive (z_3). You brought along three outfits:
one that shrinks when wet (a_1); one that has a raincoat, boots, and
umbrella (a_2); and one with only a raincoat (a_3). Considering the
problem of carrying rain gear around London all day and the uncom-
fortable nature of being drenched, you construct the following table
of losses of utility:

$W(\underline{\theta}, \underline{a})$	a_1	a_2	a_3
No rain (θ_1)	0	1	3
Rain (θ_2)	5	3	2

The past performance of the rain indicator is posted in your hotel room,
so you can assess the probability of observing a weather indication z_j
when θ_i is the true state of nature:

$f(z_j\|\theta_i)$	z_1	z_2	z_3
θ_1	0.60	0.25	0.15
θ_2	0.20	0.30	0.50

a.   List all of your possible strategies.
b.   Compute expected loss corresponding to each strategy and state of
nature combination.
c.   Plot the expected losses for your strategies and identify the admissible ones.
d.   After this analysis you decide that you prefer two of the strategies,
but cannot decide between them:

$\underline{s} = (a_1, a_2, a_3)$
[or
$\underline{s}(z_1) = a_1$, $\underline{s}(z_2) = a_2$, $\underline{s}(z_3) = a_3$]
$\underline{t} = (a_2, a_3, a_3)$
[or
$\underline{t}(z_1) = a_2$, $\underline{t}(z_2) = a_3$, $\underline{t}(z_3) = a_3$]

Interpret each of $\underline{s}$ and $\underline{t}$ in context.
e.   To choose between $\underline{s}$ and $\underline{t}$, you do the following: toss a fair coin
two times; if two heads occur, then you select $\underline{s}$; otherwise select $\underline{t}$.
What is your expected loss of utility with this rule?
f.   Suppose it rains in London 2/3 of the time. What is the best pure
strategy (the one that minimizes expected loss)? With a graphic
method, find the best among all strategies (pure and randomized).
Is the best strategy pure or randomized?

5.27   Jane Doe is the director of a personnel training program for a large company. She has developed three versions of the training program: a_1, a short course designed for persons needing only remedial or refresher training; a_2, a regular training program of medium length; and a_3, an intensive course designed for people with no prior experience and minimal education. Clearly, the choice of a training program and its resultant effectiveness depends largely on the abilities of the trainees, which may be a combination of education, work experience, and natural abilities. Jane decides that for this training session there are three possible states of nature: θ_1, high-quality trainees; θ_2, medium-quality trainees; and θ_3, poor-quality trainees. Her assessment of the loss of utility with each action–state of nature combination is given in the following table:

$W(\theta, a)$	a_1	a_2	a_3
θ_1	0	2	4
θ_2	5	3	5
θ_3	10	9	6

The experiment she performs is to look at the resumes of present employees and assess the relationship between the highest degree earned by the employee and the quality of their work. She arrives at the following conditional probability table, where z_1 denotes a Ph.D. degree, z_2 denotes a master's degree, z_3 denotes a bachelor's degree, and z_4 denotes a high-school diploma (these are highest degrees earned).

$f(\theta \mid z)$	z_1	z_2	z_3	z_4
θ_1	0.5	0.4	0.1	0.0
θ_2	0.2	0.5	0.2	0.1
θ_3	0.0	0.2	0.5	0.3

a.   List five strategies that Jane Doe could take in deciding on a training program, and evaluate their average losses.

b.   Which is the best of these strategies if in fact 30 percent of the trainees are of high quality and 50 percent are of medium quality?

5.5 STATISTICAL HYPOTHESIS TESTING

The use of data in making decisions is a hallmark of our information age, in which the collection, storage, and retrieval of vast amounts of sample information is no problem for the computers that are becoming commonplace at our places of work and in our homes. The question for mere humans is then: What does all this information mean for the real world (state of nature) in which we live? If we (rather complimentarily) define *statistics* as the art of learning from experience, then perhaps *probability* provides a means for translating observed data into decisions about the nature of our world.

Take the legal dictum that a defendant be tried by a jury of peers. Although in general the definition of one's peers would involve consideration of many levels of identification, let us decide that at least it should mean that the sexual composition of the jury reflects the sexual composition of the area in which the defendant resides; that is, jury selection should proceed without discrimination on the basis of sex. Now, suppose we read that a jury of twelve people consists of six men and six women. Assuming that the population of the jurisdiction is roughly half male and half female (as it is for the population as a whole), how does this sample information affect your assessment of the presence of discrimination in the jury selection procedure? We would probably want to ask other questions about these twelve folks, of course, but a sample proportion of six females out of twelve jurors is consistent with a selection procedure in which the selection probability of a female for jury duty is $1/2$; that is, $P(F|J) = P(F) = 1/2$, or sex and jury selection are independent. What if the proportion of women on that jury were only 5 of 12? Depending on our degree of legalism (and our personal attitude toward the laws of chance), we might still accept that the procedure was nondiscriminatory. But suppose that every member of a jury was male. Then, we argue, if the selection procedure is equally likely to select males as females $[P(F|J) = P(M|J) = 1/2]$ and if jury members are chosen independently, then the chance of all 12 being male would be $(1/2)(1/2) \cdots (1/2) = (1/2)^{12} = 1/4096 = 0.00024$—in short, it would take a minor miracle to explain it. Therefore, we would believe that some sort of bias had been present in the selection process.

We have just carried out what statisticans like to call a *hypothesis test*, here an investigation of a conjecture about a probability relative to a jury selection procedure. In the language of this chapter, let θ be the relative frequency with which the procedure selects females (say) for jury duty. We wish to investigate the hypothesis (conjecture) that the procedure selects males and females equally often; we write this as

$$H: \theta = \frac{1}{2}$$

Our experiment involves a sample experiment in which the random variable Z is the number of women selected in the jury of 12. Our decision about H is based on an assessment of a posterior probability $P(Z|\theta = 1/2) = P(Z|H)$, the probability of observing a particular jury composition, if $\theta = 1/2$ is the true state of nature. This probability $P(Z|H)$ can be viewed from two different standpoints. On the one hand, $P(Z|H)$ is the credibility we attach to the outcome of Z, computed on the basis of the statistical hypothesis H. On the other hand, when we actually observe the value of Z, we tend to think that H is less likely the smaller $P(Z|H)$ is. For the latter reason, we call $P(Z|H)$ the *likelihood* of the hypothesis H, in view of the fact that a particular value for Z is observed.

Now, it follows from Chapter 2 [e.g., Equation (2.2.1)] that $P(H|Z) = P(Z|H)P(H)/P(Z)$, where we may view $P(H)$ as the prior probability (or credibility) of the statistical hypothesis H before observing the value of Z, and $P(Z)$ is the probability (credibility) of the sample information Z. Pólya (1954) expresses this aspect of $P(Z|H)$ as (credibility *after* observation) equals (likelihood times credibility *before* observation) divided by (credibility of observation). Our assessment of the credibility of a statistical hypothesis, then, depends on an assessment of the likelihood $P(Z|H)$. However, Pólya (1954) warns: "The statistician may wisely restrict himself to the computation of the likelihood, but the statistician's customer may act unwisely if he neglects the other factors. He should carefully weigh $P(H)$, the credibility of the statistical hypothesis H before the event."

A statistical hypothesis test may be considered a choice between competing explanations of observed phenomena. In our jury example, an all-male jury may be explained in two ways: (1) the selection procedure is impartial and the all-male composition has occurred purely "by chance"; (2) the selection procedure is biased and chooses men for jury duty more often than women. We choose between these rival explanations by noting that the likelihood of an all male jury under the first explanation is 1/4096, while under the alternative, second explanation, the likelihood of this same composition is greater $[P(M|J) > P(F|J)]$.

The February 11, 1980, issue of the *New York Times* reported a reopened investigation of the assassination of President John F. Kennedy with the headline "Experts '95% Sure' 2 Fired Shots At JFK." On first glance, one might be tempted to dismiss this claim as an unsubstantiated bit of subjective probability. (Clearly, President Kennedy's assassination was a nonrepeatable event. How can any aspect of it be assigned a 95 percent probability?) In fact, this statement is a result of an application of the ideas of hypothesis testing to tape recordings of sounds of rifle fire from police records. Mark Weiss, a computer science professor at Queens College, New York City, testified that 10 sound waves, out of 12 on a tape recording of a Dallas policeman's radio transmission during the assassination, match those of a test rifle fired at the scene. "Weiss said the mathematical odds of any sound other than a rifle shot matching a rifle shot that closely is less than 5%." Thus, the experts rejected the hypothesis that the match was due to chance; it was too improbable to retain.

We are dealing here with an interpretation of coincidences: the coincidence that a jury is all male, or the coincidence that a recording sounds like two separate gun shots. If we decide to reject the hypothesis that the coincidence is due to chance then it must be attributed to some assignable cause. Here we must proceed cautiously. Consider the Baltimore Housing Study (Wilner, 1962), which addressed the effect of public housing on health and social attitudes. An elaborate study from 1954 to 1957 followed two groups of people, people living in the slums of Baltimore and people living in Lafayette Courts, a new public-housing facility. The study involved personal interviews, medical observations, monitoring of children's school performance, and so on. During the course of this longitudinal study, mortality rates were observed with the following results:

	Died	Survived	Total
Slums	6	3	9
Lafayette Courts	2	9	11
Total	8	12	20

At first glance, these data may argue strongly for public housing; the mortality rates for the two groups were

6/9 or 67% in the slums
2/11 or 18% in Lafayette Courts

Is the difference shown by such a result too large to be consistent with the chance explanation? John Tukey (1980) cautions, "Finding the question is often more important than finding the answer," and we have not quite found the right question here. Our use of the terms "such a result," "too large," and "chance" is not precise. Let us formulate the problem as follows.

You and I play a game with 20 cards of which 8 are black and 12 are red. The cards are dealt so that you receive 9 cards and I receive 11 cards. What is the probability that I receive 2 or fewer black cards?

This statement of the problem expresses the Baltimore Housing Study data in terms of a game of chance and focuses the conjecture we are to examine: the difference between slums and public housing does not really matter (does not really influence mortality), and the observed outcome is due solely to chance. The required probability can be calculated to be $335/8398 = 0.0399 \approx 1/25$. That is, an outcome apparently as favorable to public housing or more so will be produced by chance about once in 25 trials. The numerical evidence for public housing cannot be discounted but is not very strong.

Consider a situation in which the observed data produced a probability of 1/10,000 instead of 1/25. These results make it difficult to accept the conjecture of pure chance, but they do not *prove* the superiority of public housing. The data simply provide a strong argument for a nonrandom difference in mortality between the two kinds of people; they do not say what the nature of this difference is. If only young and healthy people are admitted to Lafayette Courts, for example, and slum residents are elderly and infirm, the argument in favor of the superiority of public housing would be weak indeed.

The concept of hypothesis testing is probably one of the most used (and misused) of all statistical applications of probability theory. We may make the analogy of a legal trial, the purpose of which is to examine the presupposition that a defendant is innocent (H). The question put to the jury may be summarized as follows: Is the evidence (sample information) more likely if we acquit (accept H) or if we convict (reject H)? Notice that a hypothesis test, as a legal trial, is not without risk. In the long run, we would hope that most of our decisions would be correct, but we run the risk of accepting a false conjecture or of rejecting a true one, in some of the cases. Of course, we try to minimize the risks of these two errors, but in general this is not always possible. Let us consider an example of quality control.

A manufacturer of software (programs) for personal computers is willing to allow errors (glitches) in only 10 percent of the programs produced. To check the production process, a quality control director randomly selects 10 programs from each day's production run and concludes that the process is running properly if at least nine programs work. With this rule, for what proportion of batches will the director correctly conclude that the process is operating satisfactorily? If θ is the true proportion of defective programs and Z is the number of defectives in the observed sample, we require an assessment of $P(Z \leq 1 | \theta = 0.1)$. Since the allowed defective program could be any 1 of the 10 examined, this probability is

$$P(Z = 0 | \theta = 0.1) + P(Z = 1 | \theta = 0.1) = (1 - 0.1)^{10} + (10)(0.1)(1 - 0.1)^9$$

$$= 0.3487 + 0.3874 = 0.7361$$

That is, if the process actually produces 10 percent defectives, about 74 percent of the batches will be accepted as satisfactory, while 26 percent will be unnecessarily rejected. We will call this 26 percent the producer's risk using the stated quality control rule.

Suppose now that the production process shifts to 20 percent defective programs; what are the chances that our rule fails to detect this? In this case,

$$P(Z \leq 1 | \theta = 0.2) = (1 - 0.2)^{10} + (10)(0.2)(1 - 0.2)^9$$

$$= 0.1074 + 0.2684 = 0.3758$$

is the probability that our rule accepts this bad batch, a consumer's risk. What happens to the risk of error with our present rule as the true values of θ vary? A graph of $P(Z \leq 1 | \theta)$ is given in Figure 5.6 as a function of θ. (Values for this probability when $\theta > 0.1$ represent consumer's risk; for $\theta \leq 0.1$, values are 1 - producer's risk, called the operating characteristic or O.C.)

This example allows us to examine an important fallacy associated with test of hypotheses. If we repeat verification of a conjecture (here that $\theta = 0.1$), we will compound our chances of rejecting it even when it is true. For n independent production batches all with $\theta = 0.1$, the probability that we reject at least one is 1 - P(all batches are accepted) = $1 - (0.74)^n$, a number we can

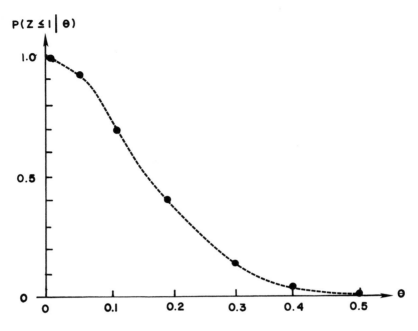

Figure 5.6   Risk of the quality control procedure.

make arbitrarily close to one if we take n large enough. In fact, this number is approximately 0.8 for n as low as 5.

How can we control the risk of accepting bad batches? One possibility is to use a stricter acceptance procedure; for example, we might accept a batch if all ten sampled programs performed satisfactorily. In Figure 5.7 we plot $P(Z = 0|\theta)$ along with our previous picture of $P(Z \leq 1|\theta)$. We have succeeded in reducing the consumer's risk $P(Z|\theta > 0.1)$ at the expense of an increase in the producer's risk $P(Z|\theta \leq 0.1)$. In the exercises for this section you are asked to investigate a solution to the problem of decreasing both error risks simultaneously.

The point of this concluding section is to demonstrate that chance is always a possible conjecture to explain an observed phenomenon. The appropriate procedure in testing this conjecture (as in a court of law) is to assume chance (innocence) until there is definite evidence to the contrary. Pólya (1954, pp. 76–77) summarizes:

> The actual occurrence of an event to which a certain statistical hypothesis attributes a small probability is an argument against that hypothesis, and the smaller the probability, the stronger is the argument. . . . We described

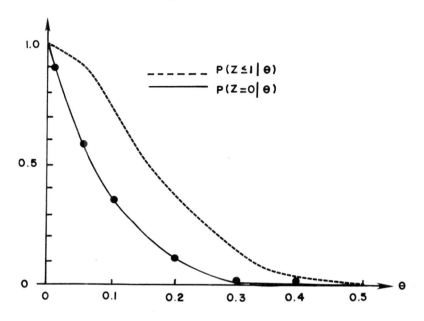

Figure 5.7 Comparison of risks of two quality control rules.

certain circumstances under which we can reasonably consider a statistical hypothesis as practically refuted by the observations. On the other hand, if a statistical hypothesis survives several opportunities of refutation, we may consider it corroborated to a certain extent.

This discussion leads us quite naturally to a study of statistical inference, a central topic of modern statistics and data analysis. We encourage you to proceed with this study, using one of the many excellent texts on this topic.

Exercises

5.28 In a recent year 40,500 babies in the U.S. died before they were 28 days old. Of those, 30,000 were white and 10,500 were nonwhite. Comment on the statement: Black kids have a better chance to survive in America than white kids. (Source of Data: *1981 U.S. Fact Book, The Statistical Abstract of the U.S.* Numbers rounded to nearest 500.)

5.29 You arrive on campus for a class and discover that you have left your watch at home. Your university has clocks on three towers, and you see that two clocks out of the three show the same time. Can you rely on the time they show? (Hint: What is the probability that two of the three clocks will agree merely by chance?)

5.30   [Based on Pólya (1954).] In the situation of Exercise 5.29, suppose two of the clocks are less than two minutes apart. Can you rely on the time in this case?

5.31   Consider a quality-control scheme to investigate θ, the percentage of defective items in a large batch. A sample of five is selected and one defective item is observed. Show that a decision taking $\theta = 1/5$ maximizes the likelihood among possible values (0, 1/5, 2/5, 3/5, 4/5, 1).

5.32   For the quality control procedure discussed in this section, calculate the risks of the original acceptance rule ($Z \leqslant 1$) with a sample of size 15. Comment on the effect of sample size on producer's and consumer's risks.

5.33   In an ESP experiment, a die is rolled and the subject tries to make it land showing 6 spots. This is repeated 720 times and the die lands 6 in 150 of these trials. Does this prove ESP exists?

5.34   Graduate admissions records show the following results for the largest graduate department at a major state university:

	Men	Women
Admit	509	89
Deny	316	19

Did this department discriminate against the men?

5.35   In a recent survey in Austin, Texas, six residents were questioned on their opinion of an expressway extension that would adversely affect a city park containing a natural spring. Of the six people, five were in favor of the extension and one was opposed.

    a.   Using the relative frequency interpretation of probability, what is the risk of error if we reject the null hypothesis that residents of Austin are ambivalent on this issue (i.e., 50 percent of the people would oppose)?

    b.   Suppose, in fact, the residents are 2 to 1 in favor of the extension. What is the risk with which we accept the null hypothesis of indifference?

5.36   [Based on Joshi (1982).] An inventory of 100 computer programs is known to be of one of the following compositions.

    a.   99 correct programs, 1 defective

    b.   2 correct and 98 defective

    c.   1 correct and 99 defective

One computer program is selected at random and carefully tested. Consider the following two strategies.

    i.   If the tested program is correct, choose between compositions a and b with probabilities 100/199 and 99/199, respectively. If the tested program is defective, choose between c and b with the same respective probabilities.

ii. If the tested program is correct, choose a. If the program is defective, choose between b and c with probabilities 99/197 and 98/197, respectively.

Suppose that the loss for an incorrect decision is $100.00. Show that the expected losses under the two strategies imply that the first is preferable using the minimax criterion. Explain why this strategy is unreasonable!

appendix A

PARADOXES OF PROBABILITY

A.1 NONTRANSITIVITY PARADOX

If Travis is taller than Emily and Emily is taller than Lucy, then we are safe in concluding that Travis is taller than Lucy. This is a result of the fact that "is taller than" is an example of a transitive relation. In general, a relation R is *transitive* if x R y and y R z together imply that x R z. Many mathematical relations are transitive: "greater than," "less than," "is isomorphic to," and "equals." So too are other relations, such as "earlier than" and "inside of." Intuitively, one feels that relations having to do with dominance, as "is better than" or "wins at tennis from" should be transitive; but not all of them are, as we saw in the menu decision problem of Chapter 5, and as we see in tennis matches when a master player is beaten by a second-rate player who has been beaten repeatedly by many of the players that the master beats. Here is a case where A (master) beats B and B beats C, but C beats A. This section investigates the implication of such cases in practical situations involving uncertainty.

Recall that a discrete random variable is a function (denoted by a capital letter, as X) which takes on values with well-defined or properly assessed probabilities. To describe a discrete random variable, it is enough to present its distribution, i.e., to indicate its possible values and corresponding probabilities. For example, the statements $P(X = 4) = 0.6$ and $P(X = 3) = 0.4$ describe a random variable X taking the values 3 and 4 with probabilities of 0.6 and 0.4, respectively. Alternative description of this random variable is in terms of the array

$$X = \begin{pmatrix} 3 & 4 \\ 0.6 & 0.4 \end{pmatrix} \tag{A.1.1}$$

in which the first row gives values of the random variable and the second row gives corresponding probabilities.

Let us now ask how to order or compare random variables, or specifically how to define a "less than" relation among discrete random variables. Consider, for example, the random variable X defined in Equation (A.1.1) and another random variable Y defined by

$$Y = \begin{pmatrix} 2 & 4 \\ 0.3 & 0.7 \end{pmatrix} \qquad\qquad (A.1.2)$$

Which one is "smaller"? This question is ambiguous, since if X = 3 (with probability 0.6) and Y = 2 (w.p. 0.3) then Y is smaller than X; however, if Y = 4 (w.p. 0.7) and X = 3, then X is smaller than Y. One possibility is to specify that X < Y (the random variable X is smaller than the random variable Y) if P(X < Y) > 1/2; i.e., X < Y if chances are better than 1/2 that X < Y, or if X is more often smaller than Y than vice versa. Similarly, we define X > Y (X is greater than Y) if P(X > Y) > 1/2. Note that the statement P(X < Y) < 1/2 is equivalent to the statement P(X ⩾ Y) > 1/2, which means that X is greater than Y provided that the probability of a tie (X = Y) is zero.

Returning to the variables X and Y defined above, we have

$$P(X < Y) = P(X = 3 \text{ and } Y = 4)$$

$$= P(X = 3)\,P(Y = 4)$$

$$= (0.6)(0.7) = 0.42$$

(assuming here the independence of the variables), so that P(X > Y) = 0.58 (in the absence of ties). Thus by our definition, X is greater than Y.

Now let us consider three independent, discrete random variables X, Y, and Z such that P(X = 3) = 1, P(Y = 1) = 0.4, P(Y = 4) = 0.6, P(Z = 2) = 0.6, and P(Z = 5) = 0.4. Here, X is an example of a *degenerate* random variable. Now P(X < Y) = 0.6 (X < Y) and P(Y < Z) = 0.64 (hence Y < Z) but P(Z < X) = 0.6 (Z < X). This is an illustration of the *nontransitivity paradox* of probability. If the variables are not independent, we must consider their joint distribution: P(X = 1, Y = 2, Z = 3) = 1/3, P(X = 3, Y = 1, Z = 2) = 1/3, and P(X = 2, Y = 3, Z = 1) = 1/3 is an example in which each of X, Y, and Z takes on one value 1, 2, or 3. This joint probability distribution could model the distribution of rank order of three candidates as preferred by the voters: a third of the people prefer candidates X, Y, and Z in that order, a third rank them as YZX, and the remaining third rank them as ZXY. In this case, P(X < Y) = 2/3, P(Y < Z) = 2/3, but P(Z < X) = 2/3; that is 2/3 of the voters prefer X to Y, 2/3 prefer Y to Z, and 2/3 prefer Z to X. If X were to run against Y, X would win; if Y were to run against Z, Y would win; but if Z were to run against X, Z would win! Substituting "proposals" for "candidates," we observe that a party in power (with the majority of votes) can often democratically impose almost any decision simply by manipulating the pairs of proposals on the ballot!

Although this paradox was first recognized in France in 1785 by the Marquis de Condorcet (1743-1794) and rediscovered by Charles Dodgson [Lewis Carroll (1832-1898)], mathematician and author of *Alice in Wonderland*, it was not fully understood by political theorists until the mid-1940s, when D. Black, a Welsh economist, rediscovered it in his work on committee decision making (1958).

The voting paradox arises in any ordinary situation in which a decision must be made between two alternatives from a set of three or more: three men simultaneously propose marriage to a woman, an applicant has a choice of three jobs, a family has three vacation choices, and so on. An example involving a choice between three pies has been suggested by P. Halmos (see Gardner, 1974). A café lists on its menu apple pie, blueberry pie, and cherry pie, but only two of them are available at any given meal. The customer may rank the pies with respect to taste, freshness, and size, and it is quite possible that the customer will prefer apple to blueberry, blueberry to cherry, and cherry to apple.

Nontransitive orderings are quite common in sports (especially in round-robin tournaments between teams) and in various games of chance. One of the first examples was given by B. Efron as quoted in Gardner (1970), using a game between two players with four dice, A, B, C, D, portrayed below.

A	B	C	D
0	3	2	5
4 0 4	3 3 3	2 2 2	1 1 1
4	3	6	5
4	3	6	5

Note that the constant (or degenerate) die B is analogous to the degenerate random variable X defined above. You allow your opponent to choose any die from this set, and then you select one of the remaining three. Both dice are tossed and the higher number wins. Since your opponent is allowed to choose first, he or she has the opportunity to choose the best die, so the game must be fair or favor your opponent. However, regardless of which die the opponent picks, you can always pick a die that has a 2/3 probability of winning, i.e., one giving 2:1 odds in your favor. In terms of the distribution of the random variables involved, we have

$$P(A = 0) = 1/3 \qquad P(A = 4) = 2/3$$
$$P(B = 3) = 1$$
$$P(C = 2) = 2/3 \qquad P(C = 6) = 1/3$$
$$P(D = 1) = 1/2 \qquad P(D = 5) = 1/2$$

and $P(A > B) = 2/3$, $P(B > C) = 2/3$, $P(C > D) = 2/3$, but $P(D > A) = 2/3$. In other words, die A beats B, B beats C, C beats D, and D beats A. Hence, if your opponent chooses A, you should choose D; if he or she chooses B, you should choose A; if he or she chooses C, you should choose B; and if he or she chooses D, you should choose C.

An example of the nontransitivity paradox in sports is found in the famous Ali-Foreman fight in Zaire in October 1974. Ali was a 4:1 underdog; moreover, since Foreman had recently beaten Frazier and Frazier had beaten Ali, then Foreman should beat Ali. However, if we rank the three fighters on skills based on press reports, we obtain a nontransitive ranking as follows:

	Ali	Frazier	Foreman
Speed	2	1	3
Power	3	2	1
Technique	1	3	2

The ranking indicates that in at least two skills Ali should beat Foreman; this was the outcome during the actual match. See Exercise A.10 for an alternative definition of the ordering of random variables.

A.2 PAIRWISE-WORST-BEST PARADOX

A related paradox of probability which has applications in sports contests, taste preferences, voting patterns, gambling, and statistics is the *pairwise-worst-best paradox*. Recall Halmos' example of the three pies; let X, Y, and Z be random variables representing the satisfaction provided for the customer by the apple, blueberry, and cherry pies, respectively. Therefore, it is possible that confronted with apple and blueberry only, the customer will usually choose apple; but if the menu also contains cherry pie one day, the rational (although seemingly paradoxical) reasoning may well lead the customer to choose blueberry instead of apple. In speed contests this paradox shows that a race in heats is very different from a single race in which all the runners compete together; the runner most likely to win the former can be least likely to win the latter. In general the pairwise-worst-best paradox describes a situation in which pairwise comparisons of three items are consistent but counterintuitive results are reached when comparing three items together. Let us consider some definite examples.

Example A.2.1 For independent random variables X, Y, and Z, let $P(X = 4) = 1$, $P(Y = 1) = 0.22$, $P(Y = 3) = 0.22$, $P(Y = 5) = 0.56$, $P(Z = 2) = 0.49$, and $P(Z = 6) = 0.51$. Then

$$P(X < Y) = 0.56$$

$$P(X < Z) = 0.51$$

$$P(Y < Z) = P(Y = 1 \text{ and } Z \text{ has any value}) + P(Y = 3 \text{ and } Z = 6)$$
$$+ P(Y = 5 \text{ and } Z = 6) = 0.6178$$

The conclusion from these relations is that Z is the largest. However,

$$P(X = \min(X, Y, Z)) = P(X = 4, Y = 5, Z = 6) = 0.2856$$

$$P(Y = \min(X, Y, Z)) = 0.3322$$

while

$$P(Z = \min(X, Y, Z)) = 0.3822$$

So Z is most likely to be the smallest among the three variables!

Example A.2.2 [Based on an example of Blyth (1972a).] Let dependent random variables X, Y, and Z be defined by a joint distribution

$$P(X = 1, Y = 2, Z = 3) = 0.25$$

$$P(X = 2, Y = 1, Z = 3) = 0.35$$

$$P(X = 2, Y = 3, Z = 1) = 0.40$$

In this case,

$$P(X < Y) = 0.65, \qquad P(X < Z) = 0.6, \qquad P(Y < Z) = 0.6$$

which shows consistently that Z is the largest, while direct observation shows that

$$P(X = \min(X, Y, Z)) = 0.25$$

$$P(Y = \min(X, Y, Z)) = 0.35$$

$$P(Z = \min(X, Y, Z)) = 0.40$$

which indicates that Z is also most likely to be the smallest.

Hence in beauty and personality contests, such as the Miss America contest, it is not sufficient to merely rate the five finalists together; the proper procedure should be the comparison of them pairwise (there are 10 different pairs) and to verify the absence of the nontransitivity paradox; similarly, comparisons should be done for all possible triples, all possible quadrapules, and only finally for all the five finalists together. Clearly, there is only a small probability that all these comparisons will be consistent, namely that neither the nontransitivity nor the pairwise-worst-best paradoxes occur. This is merely a reflection of the fact that it is highly improbable that there is indeed a most beautiful girl.

A.3 CLOCKING PARADOX

In clocking two runners X and Y, it may be that while runner X has a better chance of breaking each possible time, runner Y is almost sure to win the race! In surveys of preference, we may ask the voters to score two candidates according to some scale. It may happen that candidate 1 has better scores in the sense of more scores greater than a for every a (say, more score him likable, better than outstanding, etc.), while nearly all voters score him lower than candidate 2 when the candidates are compared. Hence, scoring can be unreliable in some instances. These are both examples of the *clocking paradox*, which states that given two random variables X and Y, it is possible to have $P(X > Y)$ close to 1 (greater than $1/2$) but to have X be stochastically smaller than Y; i.e., $P(X \leq a) > P(Y \leq a)$, or $P(X > a) < P(Y > a)$ for any number a. Now, $P(X > Y) > 1/2$ means that with a very high probability X is greater than Y, so it would seem that it is easier for X to be greater than a than for Y to be; however, $P(X \leq a) > P(Y \leq a)$ for any number a in spite of the fact that X is larger than Y with a very high probability.

Another interpretation of the clocking paradox is that there can be a contradiction and discrepancy between two rational and seemingly reasonable definitions of "smaller than" (or "larger than") for random variables. Indeed, it is quite reasonable to assume that if X is larger than Y, then we should have $P(X > Y) > 1/2$. Similarly, it is reasonable to stipulate that if X is larger than Y, then for any real number a, $P(X > a) > P(Y > a)$. The clocking paradox shows that one definition does not necessarily imply the other, which shows in particular that our original definition of "larger than" is somewhat less than perfect.

A.4 SIMPSON'S PARADOX

The paradoxes discussed in this appendix are not, of course, paradoxes in the sense of contradictions, but paradoxes in the wider sense of counterintuitive results that make nonsense of the naive attempts (by John Stuart Mill and others) to define the meaning of "confirming instance." The crux of the matter for the three paradoxes presented thus far is that we cannot always meaningfully compare random variables without stumbling upon some sort of inconsistency. *Simpson's paradox* involves probabilities only of events and not of random variables.

In a recent investigation (see Bickel, Hammel, and O'Connell, 1975) into possible sex bias in graduate admissions at the University of California at Berkeley, Simpson's paradox reared its head. In the fall semester of 1973, the university admitted 44 percent of all male applicants and only 35 percent of female applicants. In an attempt to discover the departments "guilty" of discrimination, admission rates for each sex were tabulated by department.

Surprisingly, most departments had similar admission rates for each sex, and where there was a significant difference, it was in favor of the women! This seeming contradiction between "global" and "local" statistics is an example of Simpson's paradox, mentioned by Cohen and Nagel in 1934 but emphasized by Simpson in 1951. Stated formally, for three events A, B, and C, it is possible to have

$$P(A|B) < P(A|\bar{B})$$

and at the same time have

$$P(A|B \cap C) > P(A|\bar{B} \cap C)$$

and

$$P(A|B \cap \bar{C}) > P(A|\bar{B} \cap \bar{C})$$

In other words, if $P(A|B)$ measures the desirability of a particular method of treatment and $P(A|\bar{B})$ measures desirability for an alternate method, then the first treatment may appear better than the second, while knowing that an event C occurred makes the first treatment worse than the second and knowing that C did not occur also makes the first treatment worse than the second. Lindley and Novick (1981) show that the paradox can arise when A and B are positively associated only if C is positively associated with both A and B.

Using more formal reasoning we observe that $P(A|B)$ is an average of $P(A|B \cap C)$ and $P(A|B \cap \bar{C})$. Similarly, $P(A|\bar{B})$ is an average of $P(A|\bar{B} \cap C)$ and $P(A|\bar{B} \cap \bar{C})$. Now by assumption the components of the first average are both larger than the components of the second average, but the first average is smaller than the second. Mathematically the explanation is evident: the weights corresponding to each one of the averages (by which we multiply the corresponding components) are different, which allows for the observed seemingly contradictory phenomenon. More precisely, from the formula of total probability we have

$$P(A|B) = P(C|B)P(A|B \cap C) + P(\bar{C}|B)P(A|B \cap \bar{C})$$

[the weights being $P(C|B)$ and $P(\bar{C}|B)$] while

$$P(A|\bar{B}) = P(C|\bar{B})P(A|\bar{B} \cap \bar{C}) + P(\bar{C}|\bar{B})P(A|\bar{B} \cap \bar{C})$$

[the weights being $P(C|\bar{B})$ and $P(\bar{C}|\bar{B})$]. Observe that unless the events B and C are independent, there is no relation between values of $P(C|B)$ and $P(C|\bar{B})$ or between $P(\bar{C}|B)$ and $P(\bar{C}|\bar{B})$.

Indeed, in the example to be considered below [based on Blyth (1972b)] we will have

$P(A|B)$	= 0.11	$P(A|\bar{B})$	= 0.46
$P(A|B \cap C)$	= 0.10	$P(A|\bar{B} \cap C)$	= 0.05
$P(A|\bar{B} \cap C)$	= 0.95	$P(A|\bar{B} \cap \bar{C})$	= 0.50

(these values satisfy the Simpson's paradox) and the corresponding weights will be

$$P(C|B) = 0.99 \qquad P(\overline{C}|B) = 0.01$$
$$P(C|\overline{B}) = 0.09 \qquad P(\overline{C}|\overline{B}) = 0.91$$

[of course, $P(C|B) + P(\overline{C}|B) = P(C|\overline{B}) + P(\overline{C}|\overline{B}) = 1$], which will indeed yield the numerical equalities

$$0.11 = (0.99)(0.10) + (0.01)(0.95)$$
$$0.46 = (0.09)(0.05) + (0.91)(0.50)$$

Observe that in the first equality the weight corresponding to the small number is very large, while the weight corresponding to the large number is very small; this tends to decrease the average. Conversely, in the second inequality the weight corresponding to the small number is small, while the one corresponding to the large number is large; this tends to increase the average.

Also observe that if the events C and B were independent then

$$P(C) = P(C|B) = P(C|\overline{B})$$
$$P(\overline{C}) = P(\overline{C}|B) = P(\overline{C}|\overline{B})$$

The weights would have been the same in both equations, and the paradox would not have occurred. Indeed, in our example the events B and C are strongly positively dependent.

The practical implications of this paradox of weighted averages are seen in the following example due to Blyth (1972a), which warns against amalgamation (combination) of multivariate tables which may lead to seriously misleading results. Consider a country doctor with two types of patients: C, local ones which are the great majority, and $\overline{C}$, a few patients in the city. The doctor wishes to test a new treatment B against the old one $\overline{B}$ and assigns the new treatment mostly to his local patients with $P(B|C) = 0.91$; i.e., roughly 91 percent of local patients receive the new treatment B. This gives $P(\overline{B}|C) = 0.09$, so that 9 percent of local patients receive the old treatment. In the city, which the doctor visits only once a week, most patients receive the old treatment. More precisely, $P(\overline{B}|\overline{C}) = 0.99$ and $P(B|\overline{C}) = 0.01$. After a period of time, the doctor accumulates the data shown in Table A.1. Even a cursory scanning of these data indicates that the new treatment is inferior. Indeed, 11,000 patients received the old treatment and 46 percent of them were cured $[P(A|\overline{B}) = 0.46]$, while out of 10,100 patients receiving the new treatment only 11 percent recovered $[P(A|B) = 0.11]$.

Consider, however, the more detailed Table A.2, which gives the same information separately for local and city patients. Here we observe that in cases both of local and of city patients the new treatment is definitely superior.

Table A.1 The Country Doctor's Treatments and Results

	Treatment		
Results	Standard ($\overline{B}$)	New (B)	Totals
Not cured ($\overline{A}$)	5,950	9,005	14,955
Cured (A)	5,050 (46%)	1,095 (11%)	6,145
Totals	11,000	10,100	21,100

Indeed the recovery rate for the new treatment in the case of local patients was 10 percent [i.e., $P(A|B \cap C) = 0.10$], while for the standard treatment it was only 5 percent [i.e., $P(A|\overline{B} \cap C) = 0.05$). Similarly, in the case of the city patients the new treatment gave a 95 percent recovery rate [i.e., $P(A|B \cap \overline{C}) = 0.95$], while for the standard treatment it was only 50 percent [i.e., $P(A|\overline{B} \cap \overline{C}) = 0.50$]. The physical explanation of this phenomenon is very simple: the new treatment was applied mainly in the local (C) patients who were seriously ill (their recovery rate is at most 10 percent), while the old treatment was applied almost exclusively to city patients who were much healthier to begin with (their recovery rate is between 50 and 95 percent). The assignment was unfair as far as the new treatment is concerned; under these difficult conditions it performed very well, as indicated by Table A.2. There is strong positive dependence between B and C, while C happens to contain mostly critically ill patients. Table A.1 is seriously misleading.

To clarify this point even further, let us compute the number of (hypothetical) cures if all the patients were to receive the new treatment. From Table A.2, this is $(950 + 50 + 9000 + 1000)(0.10) + (5000 + 5000 + 95 + 5)$ $(0.95) \approx 10,700$ recoveries, while the actual number of recoveries is only $50 + 1000 + 5000 + 95 = 6145$. Conversely the reader should verify that if all received the old treatment, the number of cures would have been only 5600.

Table A.2 The Country Doctor's Treatments and Results by Type of Patient

	Treatment			
	C patients only		$\overline{C}$ patients only	
Results	Standard ($\overline{B}$)	New (B)	Standard ($\overline{B}$)	New (B)
Not cured ($\overline{A}$)	950	9000	5000	5
Cured (A)	50 (5%)	1000 (10%)	5000 (50%)	95 (95%)

From Tables A.1 and A.2 one can easily compute the weights $P(C|B)$ and $P(C|\bar{B})$. The other two weights can easily be obtained using the formulas $P(\bar{C}|B) = 1 - P(C|B)$ and $P(\bar{C}|\bar{B}) = 1 - P(C|\bar{B})$.

One way to find $P(C|B)$ is to use the formula

$$P(C|B) = \frac{P(C \cap B)}{P(B)} = \frac{P(C)\,P(B|C)}{P(B)} = \frac{(11{,}000/21{,}100)(0.91)}{10{,}100/21{,}100} = 0.99$$

and similarly for

$$P(C|\bar{B}) = P(C)\,\frac{P(\bar{B}|C)}{P(\bar{B})} = \frac{(11{,}000/21{,}000)(0.09)}{11{,}000/21{,}000} = 0.09$$

and we thus have the equations

$$P(A|B) = P(C|B)P(A|B \cap C) + P(\bar{C}|B)P(A|B \cap \bar{C})$$

$$P(A|\bar{B}) = P(C|\bar{B})P(A|\bar{B} \cap C) + P(\bar{C}|\bar{B})P(A|\bar{B} \cap \bar{C})$$

and their numerical equivalents

$$0.11 = (0.99)(0.10) + (0.01)(0.95)$$

$$0.46 = (0.10)(0.05) + (0.90)(0.50)$$

We emphasize that this paradox could not have occurred had B and C been independent; i.e., if the proportion receiving the new treatment were the same for local as for city patients. An extreme example of Simpson's paradox is the following. Suppose that $P(A|B \cap C) > P(A|\bar{B} \cap C)$ and $P(A|B \cap \bar{C}) > P(A|\bar{B} \cap \bar{C})$ while $P(A|B) \approx 0$ and $P(A|\bar{B}) \approx 1$. Consider Tables A.3 and A.4. Here, $P(A|B) = 0.1$, $P(A|\bar{B}) = 0.9$, $P(A|B \cap C) = 0.09$, $P(A|\bar{B} \cap C) = 0.08$, $P(A|B \cap \bar{C}) = 0.92$, and $P(A|\bar{B} \cap \bar{C}) = 0.91$. The same type of paradox may arise if we consider a situation related to preferences for two candidates in rural and urban areas. Note, however, that as far as the election of a candidate is concerned, the joint (misleading) table is the one which matters.

Table A.3   Alternative Data for Table A.2

Results	Treatment			
	C patients only		$\bar{C}$ patients only	
	Standard ($\bar{B}$)	New (B)	Standard ($\bar{B}$)	New (B)
Not cured ($\bar{A}$)	100	10,000	1,000	9
Cured (A)	9 (8%)	1,000 (9%)	10,000 (91%)	100 (92%)

Table A.4 Alternative Data for Table A.1

Results	Treatment (all patients)	
	Standard ($\overline{B}$)	New (B)
Not cured ($\overline{A}$)	1,100	10,009
Cured (A)	10,009 (90%)	1,100 (10%)

Far from being a contrived textbook example, Simpson's paradox occurs rather frequently in ordinary descriptive data analysis. Wagner (1982) presents two examples that show how easily the paradox can occur. In early 1979 the publishers of the magazine *American History Illustrated* were pleased to note an increase in renewal rates, from 51.2 percent in January to 64.1 percent in February. To identify kinds of subscriptions responsible for the increase, rates were broken down according to source in the following table:

Expiring Subscriptions, Renewals, and Renewal Rates, by Month and Subscription Category

Month	Source of current subscription					
	Gift	Previous renewal	Direct mail	Subscription service	Catalog agent	Overall
January						
Total	3,594	18,364	2,986	20,862	149	45,955
Renewals	2,918	14,488	1,783	4,343	13	23,545
Rate	.812	.789	.597	.208	.087	.512
February						
Total	884	5,140	2,224	864	45	9,157
Renewals	704	3,907	1,134	122	2	5,869
Rate	.796	.760	.510	.141	.044	.641

Amazingly, this table shows rates in each category declined from January to February! The misleading (aggregate) increase is due to the decline in relative importance of the subscription service category (from 0.45 to 0.09), and the fact that overall renewal probability is a weighted average of the renewal probabilities for the separate categories. Here, changes in weights as well as changes in rates determine the change in overall rate.

A similar situation arose in federal personal income tax rates from 1974 to 1978 (Wagner, 1982).

Total Income and Total Tax (in thousands of dollars), and Tax Rate for Taxable Income Tax Returns, by Income Category and Year

Adjusted gross income	1974		
	Income	Tax	Tax rate
under $ 5,000	41,651,643	2,244,467	0.054
$ 5,000 to $ 9,999	146,400,740	13,646,348	0.093
$ 10,000 to $14,999	192,688,922	21,449,597	0.111
$ 15,000 to $99,999	470,010,790	75,038,230	0.160
$ 100,000 or more	29,427,152	11,311,672	0.384
Total	880,179,247	123,690,314	
Overall tax rate			0.141

Adjusted gross income	1978		
	Income	Tax	Tax rate
under $ 5,000	19,879,622	689,318	0.035
$ 5,000 to $ 9,999	122,853,315	8,819,461	0.072
$ 10,000 to $14,999	171,858,024	17,155,758	0.100
$ 15,000 to $99,999	865,037,814	137,860,951	0.159
$ 100,000 or more	62,806,159	24,051,698	0.383
Total	1,242,434,934	188,577,186	
Overall tax rate			0.152

This table shows a decrease in tax rate for every income category, yet the overall tax rate increased from 14.1 percent to 15.2 percent. Inflation increased the relative proportion of persons, and hence taxable dollars, in higher tax brackets in 1978, thus changing the weights used in calculation of the weighted average. Wagner muses, "The reader may wish to speculate about the number of legislators who fully understand the effect of Simpson's paradox even though unaware of its official name."

A.5   PITFALLS OF EQUAL PROBABILITY

In many cases, we assume that a collection of possible events are equally probable. If positive reason exists to support this, the assumption should be used without fear of contradiction or paradox. However, if our assumption is based entirely on ignorance of the events in question, we leave ourselves open to dangerous situations. A well-known example is due to Carnap (1953). Suppose we know that each ball in an urn is either blue, red, or yellow, but we know nothing about how many balls of each color are in the urn. What is the probability that the first ball chosen will be blue? Since we do not know how many balls of each color there are, the application of the equal probability principle yields the answer 1/2 and the probability that the ball is not blue is also 1/2. If the ball is not blue, it will be red or yellow, and since we know nothing about the colors of the balls, the probability that a ball is red should be 1/4. On the other hand, we may start asking the question, What is the probability that a ball is red?, and the argument as above will yield the contradictory result that P(red) = 1/4 and P(blue) = 1/2.

We can push this argument even further. What is the probability that there exists plant life on a certain distant planet outside our galaxy? Since the argument on both sides is based on ignorance, the probability in question is 1/2. What is the probability that animal life exists on the same planet? Again, the probability is 1/2. What is the probabilty that there exists either plant or animal life? Using the formula $P(A \cup B) = P(A) + P(B) - P(A \cap B)$, we may have this probability close to 1, and it is not difficult to change the example to have this probability exactly 1 (by taking two *disjoint* events such as "the hair of inhabitants on Saturn is either green or red"). This line of argument has been used by the great mathematician Blaise Pascal (1623-1662), in his essay *Pensées*, in connection with the various proofs for the existence of God.

Another pitfall of the equiprobability assumption is that it is not invariant (does not stay unchanged) under simple transformations when common sense tells us it should be. For example, if the length of a cube's edge is equally likely to be between 1 foot and 3 feet, then the probability that this edge is between 1 foot and 2 feet is 1/2. The range of edges from 2 to 3 feet corresponds to range of volume between 8 and 27 feet and the range of edges between 1 foot and 2 feet corresponds to range of volume between 1 foot and 8 feet, which is almost three time smaller. Therefore, if equal probability assumption applies to the size of the edges, it is violated when applied to volume of a cube and vice versa. Therefore, if we are told that the size of a cube (without specifying whether we refer to a length of its edge or its volume) is such that the edge of the cube is between 1 foot and 3 feet, then application of the equal probability principle must lead to a contradiction, and indeed such an application of this principle, *if we don't know how the size was selected*, is meaningless.

Exercises

A.1   Check whether the following random variables obey the nontransitivity paradox: $P(A = 3) = 1$, $P(B = 2) = 0.4$, $P(B = 4) = 0.6$, $P(C = 1) = 0.6$, $P(C = 5) = 0.4$.

A.2   Consider the following set of dice for the game of Section A.1.

A		B			C			D		
	2		0			6			4		
3	9 3	1	8 7		5	6 5		12	4 12
	10		8			6			4		
	11		8			6			4		

Giving your opponent the first choice of die and with the rule that the higher toss wins, show that you can always pick a die that has 2/3 probability of winning.

A.3   Nine tennis players are ranked in ability from 1 to 9, the best player ranked as 9 and the worst given the number 1. The nine players are divided into 3 teams indicated by the rows of the matrix

         8 1 6

         3 5 7

         4 9 2

In a round-robin tournament between teams each member of one team plays once against each member of the others. Analyze the situation with respect to the nontransitivity paradox.

A.4   Consider the following dependent variables X, Y, and Z: $P(X = 1, Y = 2, Z = 3) = 0.3$, $P(X = 2, Y = 1, Z = 3) = 0.4$, $P(X = 3, Y = 2, Z = 1) = 0.3$. Write explicitly the probabilities leading to a pairwise-worst-best paradox.

A.5   Show that one cannot construct an example of the clocking paradox for independent random variables. [Hint: This exercise involves continuous probability theory. Show that if the variables are independent, then $P(X \leq a) > P(Y \leq a)$ must imply that $P(X \leq Y) \geq 1/2$.]

A.6   Let A denote "patient is male," B denote "standard treatment for a disease," and C denote "patient cured of disease." Analyze the following tables from the point of view of Simpson's paradox.

	Males (A)				Females (A)		
	B	$\bar{B}$			B	$\bar{B}$	
C	10	100	110	C	100	50	150
$\bar{C}$	100	730	830	$\bar{C}$	50	20	70
	110	830	940		150	70	220

Males and Females Together			
	B	$\bar{B}$	
C	110	150	260
Not C	150	750	900
	260	900	1160

A.7 Analyze the following three tables, and explain why together they are an example of Simpson's paradox.

Table A

| Balls | Hats | |
	Black	Gray
Red	5	3
Blue	6	4

Table B

| Balls | Hats | |
	Black	Gray
Red	6	9
Blue	3	5

Table C: Combined Data

| Balls | Hats | |
	Black	Gray
Red	11	12
Blue	9	9

A.8 Analyze the relationship between dice A, B, and C.

A	B	C
1	1	2
6 6 8	5 3 5	4 4 2
8	7	9
1	7	9

A.9 Analyze the relationship between the following sets of six dice:

D	E	F	G	H	I
0	3	2	5	4	4
4 0 4	3 3 3	2 2 2	1 1 1	4 4 4	3 3 3
4	3	6	5	4	4
4	3	6	5	4	4

A.10 Suppose we define $X > Y$ if $E(X) > E(Y)$. Can the nontransitivity paradox exist with this definition?

A.11 (Based on an exercise in *Mathematical Spectrum, 13*(3), 1980/81). An ice-cream shop sells three flavors of ice cream, and their quality is measured on a scale of 1 to 6 (with 1 being low and 6 high). The probabilities of the quality of each flavor are given by the following table:

Flavor	Quality	Probability
Banana	3	1
Chocolate	$\begin{cases} 1 \\ 5 \end{cases}$	0.52 0.48
Vanilla	$\begin{cases} 2 \\ 4 \\ 6 \end{cases}$	0.55 0.23 0.22

If you do not like vanilla, which flavor should you choose to maximize your probability of obtaining the best quality? If you like all three flavors equally, which one should you now choose?

appendix B

TABLES

Table B.1 Probabilities in Rolling Two Fair Dice

Point x	Probability in a single fair roll P(x)
2	1/36 = 0.0278
3	2/36 = 1/18 = 0.0556
4	3/36 = 1/12 = 0.0833
5	4/36 = 1/9 = 0.1111
6	5/36 = 0.1389
7	6/36 = 1/6 = 0.1667
8	5/36 = 0.1389
9	4/36 = 1/9 = 0.1111
10	3/36 = 1/12 = 0.0833
11	2/36 = 1/18 = 0.0556
12	1/36 = 0.0278

Table B.2   Rank Order Distribution of Five-Card Poker Hands

Type of hand	Number of possibilities
Royal flush	4
Straight flush	36
Four of a kind	624
Full house	3,744
Flush	5,108
Straight	10,200
Three of a kind	54,912
Two pairs	123,552
One pair	1,098,240
Bust	1,302,540
Total	2,598,960

Table B.3   Craps Bets: Odds and House Percentages

Bet	Actual odds	Casino pays	House percentage
Pass or come	18.255:17.745	1:1	1.4141
Don't pass or don't come	17.745:17.255	1:1	1.4026
4 or 10 place	2:1	9:5	6.6667
5 or 9 place	3:2	7:5	4.0000
6 or 8 place	6:5	7:6	1.5151
Pass line plus single odds	30.255:29.745	1:1 + odds	0.8485
Pass line plus double odds	42.255:41.745	1:1 + odds	0.6060
Don't pass plus single odds	29.745:29.255	1:1 + odds	0.8320
Don't pass plus double odds	41.745:41.255	1:1 + odds	0.5915
Field[a]	20:19	1:1	2.5641
Big 6 or big 8	6:5	1:1	9.0909
Any crap	8:1	7:1	11.1111
Hardways 4 or 10	8:1	7:1	11.1111
Hardways 6 or 8	10:1	9:1	9.0909
11 of 3 proposition	17:1	15:1	11.1111
2 of 12 proposition	35:1	30:1	13.8889
Any 7	5:1	4:1	16.6667

[a]The field bet pays double for 2, triple for 12, and even for 3, 4, 9, 10, and 11.

Table B.4   Values of -p log$_2$ (p)

p	0	1	2	3	4	5	6	7	8	9
0.00	.0000	.0100	.0179	.0251	.0319	.0382	.0443	.0501	.0557	.0612
0.01	.0664	.0716	.0766	.0814	.0862	.0909	.0955	.0999	.1043	.1086
0.02	.1129	.1170	.1211	.1252	.1291	.1330	.1369	.1407	.1444	.1481
0.03	.1518	.1554	.1589	.1624	.1659	.1693	.1727	.1760	.1793	.1825
0.04	.1858	.1889	.1921	.1952	.1983	.2013	.2043	.2073	.2103	.2132
0.05	.2161	.2190	.2218	.2246	.2274	.2301	.2329	.2356	.2383	.2409
0.06	.2435	.2461	.2487	.2513	.2538	.2563	.2588	.2613	.2637	.2662
0.07	.2686	.2709	.2733	.2756	.2780	.2803	.2826	.2848	.2871	.2893
0.08	.2915	.2937	.2959	.2980	.3002	.3023	.3044	.3065	.3086	.3106
0.09	.3127	.3147	.3167	.3187	.3207	.3226	.3246	.3265	.3284	.3303
0.10	.3322	.3341	.3359	.3378	.3396	.3414	.3432	.3450	.3468	.3485
0.11	.3503	.3520	.3537	.3555	.3571	.3588	.3605	.3622	.3638	.3654
0.12	.3671	.3687	.3703	.3719	.3734	.3750	.3766	.3781	.3796	.3811
0.13	.3826	.3841	.3856	.3871	.3886	.3900	.3915	.3929	.3943	.3957
0.14	.3971	.3985	.3999	.4012	.4026	.4040	.4053	.4066	.4079	.4092
0.15	.4105	.4118	.4131	.4144	.4156	.4169	.4181	.4194	.4206	.4218
0.16	.4230	.4242	.4254	.4266	.4278	.4289	.4301	.4312	.4323	.4335
0.17	.4346	.4357	.4368	.4379	.4390	.4401	.4411	.4422	.4432	.4443
0.18	.4453	.4463	.4474	.4484	.4494	.4504	.4514	.4523	.4533	.4543
0.19	.4552	.4562	.4571	.4581	.4590	.4599	.4608	.4617	.4626	.4635
0.20	.4644	.4653	.4661	.4670	.4678	.4687	.4695	.4704	.4712	.4720
0.21	.4728	.4736	.4744	.4752	.4760	.4768	.4776	.4783	.4791	.4798
0.22	.4806	.4813	.4820	.4828	.4835	.4842	.4849	.4856	.4863	.4870
0.23	.4877	.4883	.4890	.4897	.4903	.4910	.4916	.4923	.4929	.4935
0.24	.4941	.4947	.4954	.4960	.4966	.4971	.4977	.4983	.4989	.4994
0.25	.5000	.5006	.5011	.5016	.5022	.5027	.5032	.5038	.5043	.5048
0.26	.5053	.5058	.5063	.5068	.5072	.5077	.5082	.5087	.5091	.5096
0.27	.5100	.5105	.5109	.5113	.5118	.5122	.5126	.5130	.5134	.5138
0.28	.5142	.5146	.5150	.5154	.5158	.5161	.5165	.5169	.5172	.5176
0.29	.5179	.5182	.5186	.5189	.5192	.5196	.5199	.5202	.5205	.5208
0.30	.5211	.5214	.5217	.5220	.5222	.5225	.5228	.5230	.5233	.5235
0.31	.5238	.5240	.5243	.5245	.5247	.5250	.5252	.5254	.5256	.5258
0.32	.5260	.5262	.5264	.5266	.5268	.5270	.5272	.5273	.5275	.5277
0.33	.5278	.5280	.5281	.5283	.5284	.5286	.5287	.5288	.5289	.5291
0.34	.5292	.5293	.5294	.5295	.5296	.5297	.5298	.5299	.5299	.5300

Table B.4 (continued)

p	0	1	2	3	4	5	6	7	8	9
0.35	.5301	.5302	.5302	.5303	.5304	.5304	.5305	.5305	.5305	.5306
0.36	.5306	.5306	.5307	.5307	.5307	.5307	.5307	.5307	.5307	.5307
0.37	.5307	.5307	.5307	.5307	.5307	.5306	.5306	.5306	.5305	.5305
0.38	.5305	.5304	.5304	.5303	.5302	.5302	.5301	.5300	.5300	.5299
0.39	.5298	.5297	.5296	.5295	.5294	.5293	.5292	.5291	.5290	.5289
0.40	.5288	.5286	.5285	.5284	.5283	.5281	.5280	.5278	.5277	.5275
0.41	.5274	.5272	.5271	.5269	.5267	.5266	.5264	.5262	.5260	.5258
0.42	.5256	.5255	.5253	.5251	.5249	.5246	.5244	.5242	.5240	.5238
0.43	.5236	.5233	.5231	.5229	.5226	.5224	.5222	.5219	.5217	.5214
0.44	.5211	.5209	.5206	.5204	.5201	.5198	.5195	.5193	.5190	.5187
0.45	.5184	.5181	.5178	.5175	.5172	.5169	.5166	.5163	.5160	.5157
0.46	.5153	.5150	.5147	.5144	.5140	.5137	.5133	.5130	.5127	.5123
0.47	.5120	.5116	.5112	.5109	.5105	.5102	.5098	.5094	.5090	.5087
0.48	.5083	.5079	.5075	.5071	.5067	.5063	.5059	.5055	.5051	.5047
0.49	.5043	.5039	.5034	.5030	.5026	.5022	.5017	.5013	.5009	.5004
0.50	.5000	.4996	.4991	.4987	.4982	.4978	.4973	.4968	.4964	.4959
0.51	.4954	.4950	.4945	.4940	.4935	.4930	.4926	.4921	.4916	.4911
0.52	.4906	.4901	.4896	.4891	.4886	.4880	.4875	.4870	.4865	.4860
0.53	.4854	.4849	.4844	.4839	.4833	.4828	.4822	.4817	.4811	.4806
0.54	.4800	.4795	.4789	.4784	.4778	.4772	.4767	.4761	.4755	.4750
0.55	.4744	.4738	.4732	.4726	.4720	.4714	.4708	.4702	.4696	.4690
0.56	.4684	.4678	.4672	.4666	.4660	.4654	.4648	.4641	.4635	.4629
0.57	.4623	.4616	.4610	.4603	.4597	.4591	.4584	.4578	.4571	.4565
0.58	.4558	.4551	.4545	.4538	.4532	.4525	.4518	.4511	.4505	.4498
0.59	.4491	.4484	.4477	.4471	.4464	.4457	.4450	.4443	.4436	.4429
0.60	.4422	.4415	.4408	.4401	.4393	.4386	.4379	.4372	.4365	.4357
0.61	.4350	.4343	.4335	.4328	.4321	.4313	.4306	.4298	.4291	.4283
0.62	.4276	.4268	.4261	.4253	.4246	.4238	.4230	.4223	.4215	.4207
0.63	.4199	.4192	.4184	.4176	.4168	.4160	.4152	.4145	.4137	.4129
0.64	.4121	.4113	.4105	.4097	.4089	.4080	.4072	.4064	.4056	.4048
0.65	.4040	.4031	.4023	.4015	.4007	.3998	.3990	.3982	.3973	.3965
0.66	.3956	.3948	.3940	.3931	.3923	.3914	.3905	.3897	.3888	.3880
0.67	.3871	.3862	.3854	.3845	.3836	.3828	.3819	.3810	.3801	.3792
0.68	.3783	.3775	.3766	.3757	.3748	.3739	.3730	.3721	.3712	.3703
0.69	.3694	.3685	.3676	.3666	.3657	.3648	.3639	.3630	.3621	.3611

Table B.4   (continued)

p	0	1	2	3	4	5	6	7	8	9
0.70	.3602	.3593	.3583	.3574	.3565	.3555	.3546	.3537	.3527	.3518
0.71	.3508	.3499	.3489	.3480	.3470	.3460	.3451	.3441	.3432	.3422
0.72	.3412	.3403	.3393	.3383	.3373	.3364	.3354	.3344	.3334	.3324
0.73	.3314	.3305	.3295	.3285	.3275	.3265	.3255	.3245	.3235	.3225
0.74	.3215	.3204	.3194	.3184	.3174	.3164	.3154	.3144	.3133	.3123
0.75	.3113	.3102	.3092	.3082	.3072	.3061	.3051	.3040	.3030	.3020
0.76	.3009	.2999	.2988	.2978	.2967	.2956	.2946	.2935	.2925	.2914
0.77	.2903	.2893	.2882	.2871	.2861	.2850	.2839	.2828	.2818	.2807
0.78	.2796	.2785	.2774	.2763	.2752	.2741	.2731	.2720	.2709	.2698
0.79	.2687	.2676	.2665	.2653	.2642	.2631	.2620	.2609	.2598	.2587
0.80	.2575	.2564	.2553	.2542	.2530	.2519	.2508	.2497	.2485	.2474
0.81	.2462	.2451	.2440	.2428	.2417	.2405	.2394	.2382	.2371	.2359
0.82	.2348	.2336	.2325	.2313	.2301	.2290	.2278	.2266	.2255	.2243
0.83	.2231	.2219	.2208	.2196	.2184	.2172	.2160	.2149	.2137	.2125
0.84	.2113	.2101	.2089	.2077	.2065	.2053	.2041	.2029	.2017	.2005
0.85	.1993	.1981	.1969	.1957	.1944	.1932	.1920	.1908	.1896	.1884
0.86	.1871	.1859	.1847	.1834	.1822	.1810	.1797	.1785	.1773	.1760
0.87	.1748	.1736	.1723	.1711	.1698	.1686	.1673	.1661	.1648	.1636
0.88	.1623	.1610	.1598	.1585	.1572	.1560	.1547	.1534	.1522	.1509
0.89	.1496	.1484	.1471	.1458	.1445	.1432	.1420	.1407	.1394	.1381
0.90	.1368	.1355	.1342	.1329	.1316	.1303	.1290	.1277	.1264	.1251
0.91	.1238	.1225	.1212	.1199	.1186	.1173	.1159	.1146	.1133	.1120
0.92	.1107	.1093	.1080	.1067	.1054	.1040	.1027	.1014	.1000	.0987
0.93	.0974	.0960	.0947	.0933	.0920	.0907	.0893	.0880	.0866	.0853
0.94	.0839	.0826	.0812	.0798	.0785	.0771	.0758	.0744	.0730	.0717
0.95	.0703	.0689	.0676	.0662	.0648	.0634	.0621	.0607	.0593	.0579
0.96	.0565	.0552	.0538	.0524	.0510	.0496	.0482	.0468	.0454	.0440
0.97	.0426	.0412	.0398	.0384	.0370	.0356	.0342	.0328	.0314	.0300
0.98	.0286	.0271	.0257	.0243	.0229	.0215	.0201	.0186	.0172	.0158
0.99	.0144	.0129	.0115	.0101	.0086	.0072	.0058	.0043	.0029	.0014

appendix C

ANSWERS TO SELECTED EXERCISES

Chapter 1

1.1	(a) 2/3	(b) 1/2	
1.3	(a) 1/4	(b) 1/2	(c) 3/13	(d) 5/13
1.5	(a) 4/9	(b) 5/9	(c) 1/3	(d) 2/9
1.7	2/5		
1.10	(a) 1	(b) 2/3	(c) 1/3	(d) 0

1.15   0.60

1.20   17

1.21   0.95

1.23   62/2000

1.25   36

1.26   0.79

1.27   2/3

1.28   2/7

1.30   10:9; unfavorable

1.32   Unwilling

1.33   No

1.35 $65

1.38 $15

1.39

	Prob.	$
Phillies	26/156	-10
Dodgers	65/156	-15
Yankees	60/156	32
Royals	52/156	-5
Total		2

Chapter 2

2.1   2/3

2.3   (a) 0.0013 (= 1/750)   (b) Unwilling

2.4   (a) 0.03   (b) 0.27

2.6 $(2b)/(1 + b - g)$, which is $1/2$ if the two sexes are equally likely

2.7   Yes, $(0.5)(0.6) = 0.30$

2.10   0.0465

2.11 $0.0000005954 = (1/6)^8$

2.12   0.618

2.14   4

2.15   5:6 for X

2.17 $0.309 = (12/51)(1/2) + (39/52)(13/51)$

2.19 $1 - 0.64^3 = 0.74$

2.20 $P(E | A \cap B \cap D)$

2.22   0.00911

2.23   (a) 0.895   (b) 0.0893

2.25   (a) 0.05, assuming $P(A) = 0.01, P(B) = 0.1$
       (b) 0.01, assuming $P(A) = 1/10^6, P(B) = 1/10^5, P(B|A) = 1$

2.27   4:3

2.29   0.555

2.32   0.406

2.36   0.58

2.38   0, since $p_s = 1/3, \hat{\lambda} = 0$

2.41   (a) 0.154   (b) 0.026

Chapter 3

3.2   1 white ball in one box, all others in other box

3.4   No; $-17\cancel{c}$; house odds 5:1, fair odds 4½:1

3.7   101

3.11   Picnic; $p = 0.56$

3.12   (a) 2.3
       (b) Preparing 2 cakes is optimum strategy to maximize expected profit.
           The bakery can prepare 4 cakes and keep expected profit positive.

3.13   (a) $p(1-p)^3$ (b) $p(1-p)^{k-1}$ (c) $w = 1/p$

3.14 $22.50

3.15 $230

3.16 $15 - 10p > 5$

3.20   8% green, red, white, yellow; 0% blue

3.22   All bets have the same house percentage: 5.26%

3.25   5.26%

3.27 $P(\text{win}) = 949/1980$; house percentage = 1.36%; $E(\text{win}) = 43\cancel{c}$

3.29   5.556%

3.30   4.76%

3.32   (a) 11.111   (b) 11.111

3.33   3.38

3.37   7.87%

3.40   21/125

3.41   10/11

3.42   15/91

3.46   1.5%

3.49 $p_A = p_B = 5/14$; $p_C = 2/7$

3.52   All possibilities have equal expectations.

3.53   Expectations are all mR/nN.

Chapter 4

4.2 First entropy is 1.5, second is 1.522. Second is more informative.

4.3 0.03 bit. The experiment yields small amount of information since in most cases the answer is no.

4.4 3.32 bits

4.6 $H(X) = H(Y) = 0.971$; $H(Y|X) = 0.958$; $H(Y|x_1) = 0.910$. $H(Y|x_2) = 0.991$.

4.8 Let X = forecast experiment and Y = actual weather. Then $H(Y) = 0.97$; $H(Y|X) = 0.84$; $I(X; Y) = 13$. Max $I = 0.97$.

4.9 1.12 bits

4.10 0.02 bit

4.11 $H(Y|X) = 1.274$; $H(X \otimes Y) = 2.76$

4.12 0.0561 bit

4.22 Dependencies in written Arabic are strongest, over 50%. English and Portuguese, least.

Chapter 5

5.1

	$\underline{s}_1$ (a_1, a_2)	$\underline{s}_2$ (a_1, a_1)	$\underline{s}_3$ (a_2, a_1)	$\underline{s}_4$ (a_2, a_2)
θ_1	3.1	4	1.9	1
θ_2	4.0	3	4.0	5
θ_3	4.4	2	4.8	6

5.2

	$\underline{s}_1$ (a_1, a_1, a_1)	$\underline{s}_2$ (a_1, a_1, a_2)	$\underline{s}_3$ (a_1, a_2, a_1)	$\underline{s}_4$ (a_1, a_2, a_2)
θ_1	14	13.6	13.2	12.8
θ_2	15	18.5	16.0	19.5

	$\underline{s}_5$ (a_2, a_1, a_1)	$\underline{s}_6$ (a_2, a_1, a_2)	$\underline{s}_7$ (a_2, a_2, a_1)	$\underline{s}_8$ (a_2, a_2, a_2)
θ_1	11.2	10.8	10.4	10
θ_2	15.5	19.0	16.5	20

Note: $\underline{s}_1, \underline{s}_2, \underline{s}_4$, and $\underline{s}_8$ are admissible strategies.

5.3

	$\underline{s}_1$ (a_1, a_2)	$\underline{s}_2$ (a_1, a_1)	$\underline{s}_3$ (a_2, a_1)	$\underline{s}_4$ (a_2, a_2)
θ_1	0.9	0	0.1	1
θ_2	8.2	10	2.8	1

Note: $\underline{s}_1$ is inadmissable.

5.6   (a) a_1 is the minimax action.

5.7   The minimax strategy is $\underline{s}_8$.

5.8   The minimax strategy is $\underline{s}_4$.

5.11   B(drill) = 1.9; B(sell) = 0.08. Indifference point is 8/21 = 0.38.

5.12 $B(a_1)$ = 0.48; $B(a_2)$ = 0.46. She should not stop. Continue smoking for $p(\theta_1)$ less than 0.412.

5.13 $B(a_1|z_1)$ = 51/16; $B(a_2|z_1)$ = 56/16; $B(a_1|z_2)$ = 39/14; $B(a_2|z_2)$ = 64/14. The Bayesian strategy is (a_1, a_1).

5.19   If the deer you catch on the second day is not tagged, then the preserve has 3 deer; otherwise, the preserve has less than 3 deer.

5.20   HH for p between 0 and 1/2; HS for p between 1/2 and 2/3; SS for p between 2/3 and 1.

5.22   See 5.2.

5.23   Strategies

$\underline{s}_1$	$\underline{s}_2$	$\underline{s}_3$	$\underline{s}_4$	$\underline{s}_5$	$\underline{s}_6$	$\underline{s}_7$
a_1	a_1	a_1	a_1	a_1	a_1	a_1
a_1	a_1	a_1	a_2	a_2	a_2	a_3
a_1	a_2	a_3	a_1	a_2	a_3	a_1

$\underline{s}_8$	$\underline{s}_9$	$\underline{s}_{10}$	$\underline{s}_{11}$	$\underline{s}_{12}$	$\underline{s}_{13}$	$\underline{s}_{14}$
a_1	a_1	a_2	a_2	a_2	a_2	a_2
a_3	a_3	a_1	a_1	a_1	a_2	a_2
a_2	a_3	a_1	a_2	a_3	a_1	a_2

$\underline{s}_{15}$	$\underline{s}_{16}$	$\underline{s}_{17}$	$\underline{s}_{18}$	$\underline{s}_{19}$	$\underline{s}_{20}$	$\underline{s}_{21}$
a_2	a_2	a_2	a_2	a_2	a_3	a_3
a_2	a_3	a_3	a_3	a_1	a_1	a_1
a_3	a_1	a_2	a_3	a_1	a_2	a_3

$\underline{s}_{22}$	$\underline{s}_{23}$	$\underline{s}_{24}$	$\underline{s}_{25}$	$\underline{s}_{26}$	$\underline{s}_{27}$
a_3	a_3	a_3	a_3	a_3	a_3
a_2	a_2	a_2	a_3	a_3	a_3
a_1	a_2	a_3	a_1	a_2	a_3

$R(\underline{s}, \underline{\theta})$

	$\underline{s}_1$	$\underline{s}_2$	$\underline{s}_3$	$\underline{s}_4$	$\underline{s}_5$	$\underline{s}_6$
θ_1	0.00	0.15	0.45	0.25	0.40	0.70
θ_2	3.00	2.00	1.50	2.40	1.40	0.90

	$\underline{s}_7$	$\underline{s}_8$	$\underline{s}_9$	$\underline{s}_{10}$	$\underline{s}_{11}$	$\underline{s}_{12}$
θ_1	0.75	0.90	1.20	0.60	0.75	1.05
θ_2	2.10	1.10	0.60	2.60	1.60	1.10

	$\underline{s}_{13}$	$\underline{s}_{14}$	$\underline{s}_{15}$	$\underline{s}_{16}$	$\underline{s}_{17}$	$\underline{s}_{18}$
θ_1	0.85	1.00	1.30	1.35	1.50	1.80
θ_2	2.00	1.00	1.50	1.70	0.70	0.20

	$\underline{s}_{19}$	$\underline{s}_{20}$	$\underline{s}_{21}$	$\underline{s}_{22}$	$\underline{s}_{23}$	$\underline{s}_{24}$
θ_1	1.80	1.95	2.25	2.05	2.20	2.50
θ_2	2.40	1.40	0.90	1.80	0.80	0.30

	$\underline{s}_{25}$	$\underline{s}_{26}$	$\underline{s}_{27}$
θ_1	2.55	2.70	3.00
θ_2	1.50	0.50	0.00

Note: Admissible strategies are $\underline{s}_1$, $\underline{s}_2$, $\underline{s}_5$, $\underline{s}_6$, $\underline{s}_{15}$, $\underline{s}_{18}$, $\underline{s}_{27}$.

5.26 (a–c) See 5.23, adding 2 to $R(\underline{s}, \theta_2)$.

    (d) $\underline{s}$: if fair weather is indicated, wear shrinking outfit

           if wet weather, wear raincoat, boots, and umbrella

           if inconclusive, wear raincoat

      $\underline{t}$: if fair weather, wear a raincoat; otherwise, overdress

    (e)   Select $\underline{s}$ with probability $1/4$; select $\underline{t}$ with probability $3/4$. Call this randomized strategy $\underline{s}^*$. Then the expected loss of utility for $\underline{s}^*$ is

$$R(\underline{s}^*, \theta_1) = (0.7)(1/4) + (1.8)(3/4)$$

$$R(\underline{s}^*, \theta_2) = (2.9)(1/4) + (1.2)(3/4)$$

5.27 (a)

	$\underline{s}_1$ $(a_1, a_1,$ $a_1, a_1)$	$\underline{s}_2$ $(a_2, a_2,$ $a_2, a_2)$	$\underline{s}_3$ $(a_3, a_3,$ $a_3, a_3)$	$\underline{s}_4$ $(a_1, a_2,$ $a_2, a_2)$	$\underline{s}_5$ $(a_1, a_1,$ $a_2, a_2)$
θ_1	0	2	4	1	0.2
θ_2	5	3	5	3.4	4.4
θ_3	10	9	6	9	9.2

5.29 $3[1/(12 \cdot 60)] = 0.004$

5.30 Any one of the three clocks could be the one among the two agreeing clocks showing the earliest time. The probability that all three clocks show time less than 2 minutes apart or only two clocks do by chance is
$3 [2/(12 \cdot 60)]^2 + 2[2/(12 \cdot 60)] [1 - 2/(12 \cdot 60)] = 0.017$

5.31

θ	0	0.2	0.4	0.6	0.8	1.0
$P(Z \leqslant 1 \mid \theta)$	0	0.4096	0.2592	0.0768	0.0064	0

5.32 n = 10

θ	0.01	0.05	0.10	0.20	0.30	0.40	0.50
$P(Z = 0 \mid \theta)$	0.9044	0.5987	0.3487	0.1074	0.0282	0.0060	0.0010
$P(Z \leqslant 1 \mid \theta)$	0.9957	0.9139	0.7361	0.3758	0.1493	0.0464	0.0107

n = 15

θ	0.01	0.05	0.10	0.20	0.30	0.40	0.50
$P(Z = 0 \mid \theta)$	0.8601	0.4633	0.2059	0.0352	0.0047	0.0005	0.0000
$P(Z \leqslant 1 \mid \theta)$	0.9904	0.8290	0.5490	0.1671	0.0353	0.0052	0.0005

5.36 Expected losses under the two strategies are

| | Strategy | |
Composition	i	ii
a	50.25	1.00
b	50.25	50.75
c	50.25	50.75

Use of strategy i results in a loss (on the average) of $49.25, if the composition is of type a, and a saving of $0.50 in the other cases.

Appendix A

A.1 $P(A > B) = 0.4; P(A > C) = 0.6; P(B > C) = 0.6;$ thus $B > A, A > C,$ and $B > C$. There is no paradox.

A.2 $P(A > B) = 2/3; P(B > C) = 2/3; P(C > D) = 2/3; P(D > A) = 2/3$

A.3 Denote the teams corresponding to the three rows by A, B, and C, respectively. Then $P(A > B) = 5/9$; i.e., A beats B. $P(B > C) = 5/9$; i.e., B beats C. But $P(C > A) = 5/9$; i.e., C beats A. Thus the relation is nontransitive.

A.4 $P(X < Y) = 0.3; P(Y < Z) = 0.7, P(Z < X) = 0.3$. In pairwise comparison, X is the largest and Z is the smallest. However, $P(X = \max(X, Y, Z)) = 0.3$ and $P(Z = \max(X, Y, Z)) = 0.7$. So the joint comparison indicates that Z is most likely to be the largest.

A.6 $P(C \mid B) = 110/260; P(C \mid \bar{B}) = 1/6; P(C \mid B \cap A) = 10/110; P(C \mid \bar{B} \cap A) = 100/830; P(C \mid B \cap \bar{A}) = 100/150; P(C \mid \bar{B} \cap \bar{A}) = 50/70$. Now, $P(C \mid B) > P(C \mid \bar{B})$, but $P(C \mid B \cap A) < P(C \mid \bar{B} \cap A)$, and $P(C \mid B \cap \bar{A}) < P(C \mid \bar{B} \cap \bar{A})$.

A.8   The problem is identical to A.3. For example, $P(A > B) = 5/9$.

A.9   D beats E; D ties I; E beats F; E ties G; F beats D and G; G beats D; G ties E, H, and I; H beats D, E, and I; H ties G; I ties D and G.

A.10   No

POSTSCRIPT

In this work we have guided you (we hope) on a pleasant journey through the remarkable world of uncertainty by means of the concepts of modern probability theory. We have tried to emphasize that probability theory is not just another mathematical discipline, but rather "an extraordinary philosophical success story of modern times which has completely permeated both our practical and our theoretical lives" like no other mathematical concept (Hacking, 1974). We have tried to warn you about the misuses of probability theory and about the futility of expecting to completely control random phenomena (such as gambling). Our discussion of the probabilistic foundations of information theory has established the framework for your consideration of the quantitative description of information transmission and associated computer technology. We hope that we have helped you to face decisions in uncertainty with greater confidence and less anxiety. As British economist John Maynard Keynes (1883-1946) pointed out, "In actual exercise of reason, we do not wait on certainty."

If, indeed, we did catch your interest and convince you that the subject of probability deals with current issues in our complex, technological, uncertain world, then our goal has been well achieved. We have purposely omitted any discussion of the linear algebra, calculus, and higher mathematics which form the foundation for this study; we did not want to obscure the forest with the trees. Remember, however, that these mathematical tools are essential to a deeper understanding of this subject. We challenge and invite you to continue this journey with one of the numerous good books on probability theory, some of which are listed in the Bibliography.

Good luck in your investigation of uncertainty!

BIBLIOGRAPHY

Abramson, N. (1963). *Information Theory and Coding*. McGraw-Hill, New York.

Alexander, J. R. (1978/79). Probability as an aid in social research: the randomized response technique. *Mathematics Spectrum 11*:10–13.

Archer, J. A. (1976). The odds meet the great martingale. *Mathematics Teacher 64*:234–240.

Arrow, K. J. (1963). *Social Choice and Individual Values*. Second edition. Yale University Press, New Haven.

Ash, R. (1965). *Information Theory*. Interscience Publishers, New York.

Balinski, M. L., and Young, H. P. (1974). A new method for congressional apportionment. *Proceedings of the National Academy of Sciences 71*: 4602–4606.

Barnett, V. (1973). *Comparative Statistical Inference*. John Wiley & Sons, New York.

Bayes, T. (1958). An essay towards solving a problem in the doctrine of chances. *Biometrika 45*:293–315.

Bertrand, J. L. F. (1889). *Calcul des Probabilités*. Gauthier-Villars et fils, Paris.

Bickel, P. J., Hammel, E. A., and O'Connell, J. W. (1975). Sex bias in graduate admissions: data from Berkeley. *Science 187*:398–404.

Bissinger, B. H. (1980). Stochastic independence versus intuitive independence. *Two Year College Journal 11*:122–123.

Black, D. (1958). *The Theory of Committees and Elections*. Cambridge University Press, Cambridge.

Black, D., and Newing, R. A. (1951). *Committee Decisions with Complementary Valuation*. W. Hodge, London.

Blyth, C. R. (1972a). On Simpson's paradox and the sure-thing principle. *Journal of the American Statistical Association 67*:364–366.

Blyth, C. R. (1972b). Some probability paradoxes. *Journal of the American Statistical Association 67*:366–381.

Blyth, C. R. (1973). Simpson's paradox and mutually favorable events. *Journal of the American Statistical Association 68*:746.

Brams, S. J. (1976). *Paradoxes in Politics*. Free Press, New York

Brazier, G. D. (1980). What are the odds: construction competition probabilities. *Two-Year College Mathematical Journal 11*:291–292.

Brewer, K. R. W. (1981). Estimating marihuana usage using randomized response—some paradoxical findings. *Australian Journal of Statistics 23*: 139–148.

Brillouin, L. (1956). *Science and Information Theory*. Academic Press, New York.

Brown, G. G., and Rutemiller, H. C. (1973). Some probability problems concerning the game of bingo. *Mathematics Teacher 66*:403–405.

Bryson, M. C. (1973). Craps with crooked dice. *American Statistician 27*: 167–168.

Buchanan, J., and Tullock, G. (1962). *The Calculus of Consent*. University of Michigan Press, Ann Arbor.

Bunday, B. D. (1976). Some thoughts on a coin-tossing problem. *Mathematical Gazette 60*:108–114.

Butterworth, R. L. (1971). A research note on the size of winning coalitions. *American Political Science Review 65*:741–745.

Campbell, C., and Joiner, B. (1973). How to get the answer without being sure you've asked the question. *American Statistician 27*:229–231.

Carnap, R. (1953). What is probability? *Scientific American 188*:128–138.

Chernoff, H. (1962). The scoring of multiple choice questionnaires. *Annals of Mathematical Statistics 33*:375–393.

Chernoff, H., and Moses, L. E. (1959). *Elementary Decision Theory*. John Wiley & Sons, New York.

Clark, M., and MacNeil, A. (1976). Odd couples and missing cars. *Mathematics Spectrum 9*:42–46.

Cohen, M. R., and Nagel, E. (1934). *An Introduction to Logic and Scientific Method*. Harcourt Brace Jovanovich, New York.

Daston, L. J. (1980). Probabilistic expectation and rationality in classical probability theory. *Historia Mathematica 7*:234–260.

de Finetti, B. (1974). *Theory of Probability*. Vols. 1 and 2. John Wiley and Sons, London.

Devore, J. L. (1975). Estimating a population proportion using randomized responses. *Mathematics Magazine 52*:38–40.

Diaconis, P., and Zabell, S. L. (1982). Updating subjective probability. *Journal of the American Statistical Association 77*:822–830.

Dickey, J. M. (1979). Expert uncertainty and the use of subjective probability models. Inaugural lecture delivered at the University College of Wales, Aberystwyth.

Downs, A. (1957). *An Economic Theory of Democracy*. Harper & Row, New York.

Dubey, S. D. (1965). Statistical solutions of a combinatorial problem. *American Statistician 20*:30–38.

Dupont, R. (1981). PATCO and the fear of flying. *Washington Post*. August 11, 1981.

Epstein, R. A. (1977). *The Theory of Gambling and Statistical Logic*. Revised ed. Academic Press, New York.

Fairley, W., and Mosteller, F. (1974). A conversation about Collins. *University of Chicago Law Review 41*:242–253.

Falk, R., and Bar-Hillel, M. (1980). Magic possibilities of the weighted average. *Mathematics Magazine 53*:106–107.

Farquharson, R. (1969). *Theory of Voting*. Yale University Press, New Haven.

Fishburn, P. C. (1966). Decision under uncertainty: an introductory exposition. *Journal of Industrial Engineering 27*:341–343.

Folsom, R. E., Greenberg, B. G., Horvitz, D. G., Abernathy, J. R. (1973). The two alternate questions randomized response model for human surveys. *Journal of the American Statistical Association 68*:525–530.

Freund, J. E. (1965). Puzzle or paradox? *American Statistician 20*:29.

Frohlich, N. (1975). The instability of minimum winning coalitions. *American Political Science Review 69*:943–946.

Frohlich, N., and Oppenheimer, J. A. (1977). *Modern Political Economy*. Prentice-Hall, Englewood Cliffs, N. J.

Frohlich, N., Oppenheimer, J. A., and Young, O. R. (1971). *Political Leadership and Collective Goods*. Princeton University Press, Princeton.

Gangolli, R. A., and Yevisaker, D. (1967). *Discrete Probability*. Harcourt Brace Jovanovich, New York.

Gardner, M. (1970). The paradox of the nontransitive dice and the elusive principle of indifference. *Scientific American 223*:110–113.

Gardner, M. (1974). On the paradoxical situations that arise from non-transitive relation. *Scientific American 230*:120–124.

Gardner, M. (1976). On the fabric of inductive logic and some probability paradoxes. *Scientific American 235*:119–121.

Gardner, M. (1982). *Paradoxes to Puzzle and Delight*. Freeman, San Francisco.

Good, I. J., and Tideman, T. N. (1981). The relevance of imaginary alternatives. *Journal of Statistical Computing and Simulation 12*:313–315.

Goodstadt, M. S., Cook, G., and Gruson, V. (1978). The validity of reported drug use: the randomized response technique. *International Journal of Addictions 13*:359–367.

Groeneveld, R., and Meeden, G. (1975). Seven game series in sports. *Mathematics Magazine 48*:187–191.

Gudder, S. (1981). Do good hands attract? *Mathematics Magazine 54*:13–16.

Hacking, I. (1965). *Logic of Statistical Inference*. Cambridge University Press, Cambridge.

Hacking, I. (1974). Foreword to *Probability Theory: A Historical Sketch*, by L. E. Maistrov, translated and edited by S. Kotz. Academic Press, New York.

Haldeman, H. R., with DiMoria, J. (1978). *The Ends of Power*. New York Times Books, New York.

Harding, R. (1971). Collective action as an agreeable n-prisoners' dilemma. *Behavioral Science 16*:472–481.

Harding, R. (1976). Hollow victory: the minimum winning coalition. *American Political Science Review 70*:25–31.

Harris, L. (1967). Letter to the editor. *American Statistician 21*:42.

Heiny, R. L. (1981). Gambling, casinos, and game simulation. *Mathematics Teacher 74*:139–143.

Hever, G. A. (1980). A problem with dice. *Mathematics Magazine 53*:247.

Hille, J. W. (1978/79). A Bayesian look at the jury system. *Mathematics Spectrum 11*:45–47.

Hume, D. (1748). *An Enquiry Concerning Human Understanding*. Edited by Charles W. Hendel, 1955. Bobbs-Merrill, Indianapolis.

Huygens, C. I. (1920). *Oeuvres Complètes*. Vol. 14. Martinus Nijhoff, The Hague.

Jelinek, F. (1968). *Probabilistic Information Theory: Discrete and Memoryless Models*. McGraw-Hill, New York.

Johnson, B. R., and Smith, H. P. (1981). Draw-a-string: a slot machine game. *Mathematics Magazine 54*:190–194.

Jones, D. S. (1979). *Elementary Information Theory*. Clarendon Press, Oxford.

Joshi, V. M. (1982). A counter example of Hacking against the long run rule. *British Journal for the Philosophy of Science 33*:287–289.

Kahneman, D., and Tversky, A. (1982). The psychology of preferences. *Scientific American 246*:160–173.

Kauffman, A., and Faure, R. (1968). *Introduction to Operations Research*. Academic Press, New York.

Kellogg, P. J., and Kellogg, D. J. (1954). Entropy of information and the odd ball problem. *Physics Review 96*:1438–1439.

Kemeny, J. G., and Thompson, G. L. (1957). The effect of psychological attitudes on the outcomes of games. Contributions to the Theory of Games, III. *Annals of Mathematical Studies 39*:273–298.

Keynes, J. M. (1961). *A Treatise on Probability*. Third edition. Oxford University Press, Oxford.

Kinney, J. (1978). Tossing coins until all are heads. *Mathematics Magazine 51*: 184–185.

Klamkin, M. S. (1971). A probability of more heads. *Mathematics Magazine 44*:146–149.

Kolmogorov, A. N. (1956). *Foundations of the Theory of Probability*. N. Morrison, translator. Chelsea, New York.

Kotz, S. (1966). *Recent Results in Information Theory*. Methuen & Co., London.

Laporte, G., Ovellet, R., and Lefebure, F. (1980). A paradox in elementary probability theory. *Mathematics Gazette 64*:53–54.

Lesohin, M. M., Luk'yanenkov, K. F., and Piotrovskiĭ, R. G. (1982). *Introduction to Mathematical Linguistics*. Minsk (in Russian).

Lessing, R. (1974). The duration of bingo games. *Journal of Recreational Mathematics 7*:56–59.

Lin, C., and Berger, P. D. (1969). On the selling price and buying price of a lottery. *American Statistician 23*:25–26.

Lindley, D. V., and Novick, M. R. (1981). The role of exchangeability in inference. *Annals of Statistics 9*:45–58.

Litwin, S. (1967). A decision not based on expected monetary values. *American Statistician 21*:263.

Liu, P. T., Chow, L. P., and Mosley, W. H. (1975). Use of the randomized response technique with a new randomized device. *Journal of the American Statistical Association 70*:329–332.

MacMahon, B., Yen, S., Trichopoulos, D., Warren, K., and Nardi, G. (1981). Coffee and cancer of the pancreas. *New England Journal of Medicine 304*:630–633.

Maistrov, L. E. (1974). *Probability Theory: A Historical Sketch*. Samuel Kotz, translator and editor. Academic Press, New York.

Manuel, B. (1977). Counterfeit coin problems. *Mathematics Magazine 50*: 90–91.

Marks, D. (1980). *Psychology of the Psychic*. Prometheus Books, Buffalo, N.Y.

Mayo, M. (1962). Will this system beat roulette? *Mathematics Magazine 9*: 32–39.

Meyer, R. W. (1978). Everything you always wanted to know about the mathematics of sex and family planning . . . but were afraid to calculate. *MATYC Journal 43*:7–12.

Nelson, H. L. (1976/77). Analysis of game theoretic craps. *Recreational Mathematics 9*:94–99.

Niemi, R. G., and Riker, W. H. (1976). The choice of voting systems. *Scientific American 234*:21–27.

Ortel, M., and Rossi, J. (1981). An exercise involving conditional probability. *Mathematics Magazine 54*:125–128.

Pacific Reporter (1968). People vs. Collins, cited in W. Fairley and F. Mosteller (1974), A conversation about Collins. *University of Chicago Law Review 41*:242–253.

Packwood, R. (1981). As quoted in *Washington Post*, October 30, 1981.

Pólya, G. (1954). *Patterns of Plausible Inference*. Vol. 2 of *Mathematics and Plausible Reasoning*. Princeton University Press, Princeton.

Raiffa, H. (1968). *Decision Analysis*. Addison-Wesley, Reading, Mass.

Reader's Digest (1975). Vol. 107, no. 640, p. 152.

Robbins, H., and Van Ryzin, J. (1975). *Introduction to Statistics*. Science Research Associates, Chicago.

Rustagi, J. S. (1975). Lotteries. Ohio State University Technical Report, No. 119.

Rustagi, J. S. (1981). Probability structures of modern lottery games. Ohio State University Technical Report, No. 236.

Sagan, H. (1981). Markov chains in monte carlo. *Mathematics Magazine 54*: 3-10.

Sampson, A. R. (1974). Unexpected payoffs. *American Statistician 28*:76.

Savage, I. R. (1968). *Statistics = Uncertainty and Behavior.* Houghton Mifflin, Boston.

Savage, L. J. (1972). *The Foundations of Statistics.* Revised edition. Dover, New York.

Scarne, J. (1974). *Scarne's Complete Guide to Gambling.* Simon & Schuster, New York.

Selvin, S. (1975a). A problem in probability. *American Statistician 29*:67.

Selvin, S. (1975b). On the Monty Hall problem. *American Statistician 29*:134.

Servan-Screiber, J. J. (1981). *World Challenge.* Simon & Schuster, New York.

Shiflett, R. C., and Schultz, H. S. (1978). Can I expect a full set? *Mathematics Gazette 62*:262-265.

Simpson, E. H. (1951). The interpretation of interaction in contingency tables. *Journal of the Royal Statistical Society, Series B 13*:238-241.

Smith, A. V. (1968). Some probability problems in the game of "craps." *American Statistician 22*:29-30.

State lotteries: a legal sucker bet (1974). *Consumer Reports 39*:177-179.

Stein, J., ed., (1967). *Random House Dictionary of the English Language.* Unabridged. Random House, New York.

Sterrett, A. (1967). Gambling doesn't pay. *Mathematics Teacher 60*:210-214.

Tenney, R. L., and Foster, C. C. (1976). Non-transitive dominance. *Mathematics Magazine 49*:115-120.

Thompson, W. A., Jr. (1967). Economic probability. *American Statistician 21*:26-28.

Thurber, J. (1939). *Cream of Thurber.* Readers Union, London.

Tukey, J. W. (1980). We need both exploratory and confirmatory. *American Statistician 34*:23-25.

Tutubalin, V. N. (1972). *Probability Theory.* Moscow University Press, Moscow (in Russian).

U.S. Bureau of the Census (1981). *Statistical Abstract of the United States. U.S. Fact Book.* Washington, D.C.

Vargo, L. G. (1977). A nonparametric model for series competition. *Mathematics Magazine 50*:26–26.

Vasilenko, Y. K. (1975). On a problem of overestimation in control devices. *Matematika Shkole 22*:68–70 (in Russian).

Venn, J. (1888). *The Logic of Chance.* Macmillan, New York.

Wagner, C. H. (1982). Simpson's paradox in real life. *American Statistician 36*: 46–47.

Wallenius, K. T. (1980). Statistical methods in sole source contract negotiations. *UM 2*:113–123.

Warner, S. L. (1965). Randomized response: a survey technique for eliminating evasive answer bias. *Journal of the American Statistical Association 60*: 63–69.

Weiss, L. (1961). *Statistical Decision Theory.* McGraw-Hill, New York.

Wilner, D. M. (1962). *The Housing Environment and Family Life.* Johns Hopkins Press, Baltimore.

Wilson, A. N. (1970). *The Casino Gambler's Guide.* Harper & Row, New York.

Winans, C. F. (1976). The probability of winning a simple card game. *Mathematical Recreations 25*:7–11.

Wuffle, A. (1979). Mo Florina's advice to children and other subordinates. *Mathematics Magazine 52*:292–297.

Yaglom, A. M., and Yaglom, I. M. (1973). *Probability and Information.* Third edition. Nauka, Moscow (in Russian).

AUTHOR INDEX

Some additional authors are mentioned in the Bibliography only.

SUBJECT INDEX